MORE HOW DO THEY DO THAT?

Also by Caroline Sutton

How Do They Do That?
How Did They Do That?

MORE HOW DO THEY DO THAT?

Wonders of the Modern World Explained

Caroline Sutton
and
Kevin Markey

Illustrated by Tom Vincent

A Hilltown Book

QUILL
WILLIAM MORROW
New York

It is the policy of William Morrow and Company, Inc., and its imprints and affiliates, recognizing the importance of preserving what has been written, to print the books we publish on acid-free paper, and we exert our best efforts to that end.

Library of Congress Cataloging-in-Publication Data

Sutton, Caroline.
More how do they do that? / Caroline Sutton and Kevin Markey.
p. cm.
Includes bibliographical references.
ISBN 0-688-13221-9
1. Questions and answers. I. Markey, Kevin. II. Title.
AG195.M34 1992
031.02—dc20

92-22201
CIP

Printed in the United States of America

8 9 10

ACKNOWLEDGMENTS

WE WOULD LIKE TO THANK the following writers who contributed to the book: Allison Adato, Renée Bacher, Lynne Bertrand, Freda Garmaise, Denise Lovatt Martin, Jennifer Wolff, and Ariel Zeitlin.

?

Contents

MORE HOW DO THEY DO THAT?

?

How do they detect counterfeit bills?

ALTHOUGH THE AVERAGE PERSON might be slipped a counterfeit bill and accept it unawares, someone trained in detection can nearly always spot it instantly and effortlessly. Why? Because the government has made it all but impossible to precisely copy the real stuff. The distinctive markings of legitimate money are printed by steel engraved plates on unique paper, all of which gives Federal Reserve notes a visual sharpness unmatched by imitations. Counterfeit bills are detected either because they lack certain visual design details, attributable to poor draftsmanship or faulty printing techniques, or because they are printed on the wrong paper.

Even an exact-scale, line-by-line copy of a bill—and anyone with access to a photocopier can, but not legally, make such a copy—would fail to pass muster. It would come off looking flat in comparison to money printed on the official government plates. The most common method of counterfeiting is by photo-offset, whereby a printing plate is made from a photograph of a real bill. The effect is like a poster of a painting: visually similar, but obviously not the same thing. On a photo-offset counterfeit, the portrait fails to stand out distinctly from the background; the saw-tooth points of the Treasury seal are usually uneven; the scroll work around the border is often blurred; the background appears mottled.

Regardless of the level of printing sophistication achieved by a counterfeiter, a phony bill is always given away by the paper it is printed on. The paper used for U.S. currency, manufactured exclusively for the government by Crane and Company, a paper company headquartered in Dalton, Massachusetts, is easily identified by the tiny blue and red fibers woven into it, fibers which are visible to the naked eye. Such paper cannot legally be manufactured by anyone but Crane. Ersatz bills often have superficial blue and red markings, but upon close examination it can be seen that the faint squiggles are printed *on* the paper and not embedded *in* it.

The United States first took a serious look at counterfeiting in the early 1860s, when it is estimated some thirty percent of the currency in circulation was phony. The problem was that there were 1,600 state banks designing, printing, and issuing 7,000 varieties of notes, a practice that led to widespread confusion. It got to the point where no one really remembered what all the different bills were supposed to look like. In 1863 the government decided to simplify things by adopting a national currency. Far from being deterred, however, counterfeiters set about duplicating the new national currency with the same vigor that they had previously shown in faking the old state-bank notes. It soon became apparent that the only way to stop counterfeiting would be to actively suppress it, and in 1865 the United States Secret Service was established for that purpose.

The Secret Service has enjoyed great success in its efforts, but new printing technology, in the form of high-quality color copiers, is making counterfeit detection increasingly difficult. What clever counterfeiters are doing these days is bleaching out the printing from a one dollar bill and then using a color copier to give the blank bill a higher denomination make-over. Singles become tens, twenties, fifties, or hundreds; not pure profit, but a healthy return on a small initial investment. Since the bogus high-denomination bills are printed on good paper and have accurate designs, a person might not realize he has been passed one unless he is already looking for it. To make things easier for unsuspecting clerks, the government changed paper money at the end of 1991. (The last change was in 1928 when the government reduced the dimensions of bills from 3.1 inches by 7.4 inches to the current size, about 2.5

inches by 6.6 inches). The new currency has a clear polyester thread woven into the paper; this vertical thread (appearing to the left of the Treasury seal on high-denomination bills, between the seal and the portrait on singles) states the correct denomination of the bill. The money, called "denominated distinctive currency," will allow store and bank employees to check instantly a bill's legitimacy by holding it up to a light and reading the thread. The anticounterfeit bills will be phased in gradually, allowing time for the old, unmarked bills to be reclaimed and destroyed.

?

How do they keep you from registering to vote in more than one jurisdiction?

"WE WOULD HATE FOR YOU to say so," reports a Federal Election Commission official, "but basically there is no way to do it." So, now that we've spilled the beans, what's to keep you from registering to vote in every district from here to Kalamazoo? Well, a couple of things, foremost among them your unwavering honesty.

Each state of the union handles its own voter registration. A voter registration card typically requires that you state your current address, which to qualify you for the vote must lie within the district you are registering in, and also your previous address. Election officials can easily check the truth of the recorded addresses by referring to town census and tax lists. When you register after moving from one district to another in the same state, a town official asks if you were registered in your last community; if you respond positively, a postcard will be sent to your last town's civic officials, telling them to strike you from their voter registration list. A person wanting multiple registrations would first have to establish residency in a number of jurisdictions and then deny prior registrations (previous addresses are harder to conceal because of readily available information such as tax records and automobile registrations). Even so, there is a chance that town officials would send a courtesy notice to the old district anyway. Also, a few states maintain a central voter registration computer that contains the name and district of every registered voter and quickly catches duplicates.

Interstate registration fraud is a bit harder to detect. States are not legally required to notify other states when they pick up one of their voters, and only about two thirds of the states currently do so. Conceivably, a resident voter of say, California, could move to Texas and register to vote there while maintaining and denying his registration in California, thereby gaining an extra presidential ballot and the ability to vote for two governors and a large number of state and local office seekers. Of course, that person might end up paying all sorts of extra taxes as well.

Most people do not lie about their registration history, and most voter registration personnel are inclined to trust anyone whose sense of civic duty is strong enough to make him get out and vote. As the Federal Election Commission official put it, "They're so happy when people come in to register, they're not going to make it difficult."

?

How do they get ships through the Panama Canal?

A REMARKABLE FEAT of engineering, the Panama Canal covers a distance of fifty-one miles and lifts a ship as high as a nine-story

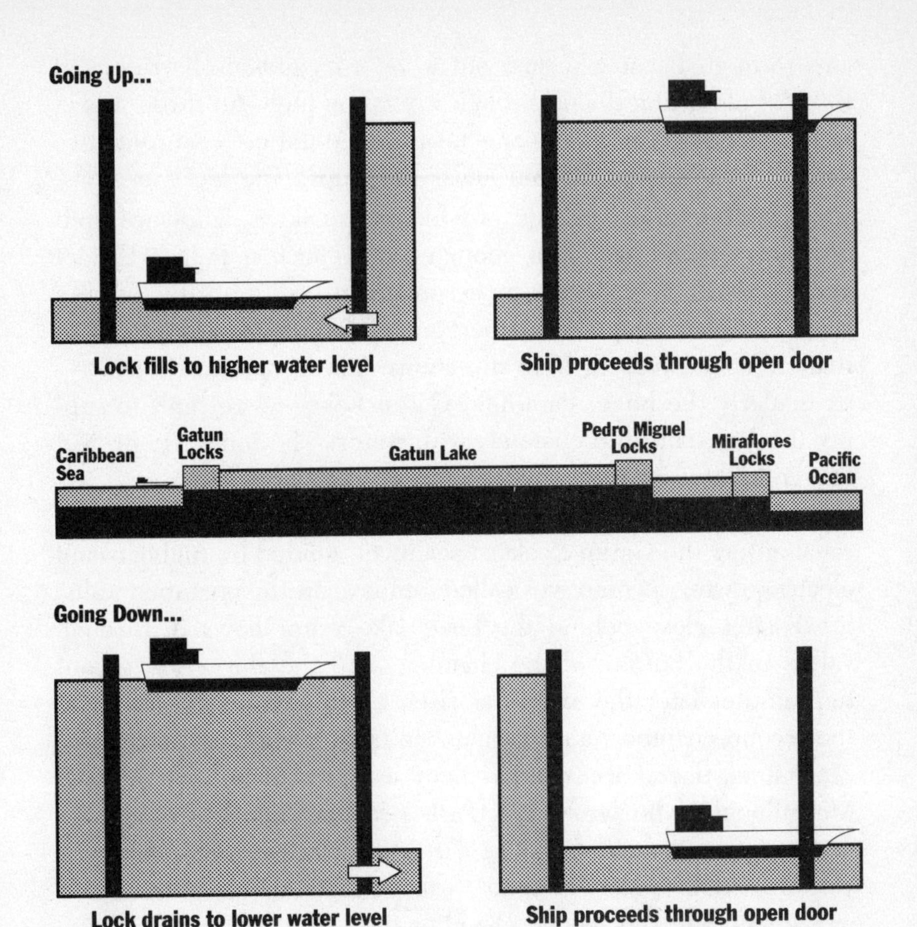

When a ship enters the Panama Canal, water flows into the lock to raise the craft to the height of a nine-story building. To passengers, the lift feels like an elevator ride. At the other end of the canal, water drains from the lock to lower the ship to sea level.

building in order to overcome the high terrain of the Isthmus of Panama. The canal was built to provide a shortcut for ships traveling from the Atlantic Ocean to the Pacific or vice versa. No longer would they have to make the arduous 7,500-mile voyage around Cape Horn, the southernmost tip of South America.

The original plan for the canal called for cutting through the isthmus with a sea-level channel, and the huge undertaking began that way in the late 1800s. But the rocky continental divide that cuts through Panama turned out to be a formidable barrier, and the cost of cutting down through it was too high. So three sets of locks were built instead—one near the Atlantic Ocean and two near the Pacific—to lift and lower the ships. The locks are massive chambers, into and out of which water flows. They are built of concrete, and are large enough—at 1,000 feet long, 110 feet wide and 70 feet deep—to accommodate all but the world's largest military ships and supertankers. The locks are paired so that ships can pass in both directions. Three man-made lakes—particularly the huge, dam-held Gatun Lake—were built to supply the locks and the channels with water. The immense project was completed in the summer of 1914.

A ship that begins at the Caribbean, or Atlantic, end of the canal enters the Gatun Locks at sea level, guided by tugboats and electric towing locomotives called "mules" on the chamber walls. Steel gates close behind the boat, lake water flows in through valves in the bottom of the chamber, and the ship rises. About ten minutes later the water has risen to the level of the water in the second chamber, and the gates in front of the boat swing out. (The steel doors are nearly silent as they move, says David McCullough, who wrote *The Path Between the Seas: The Creation of the Panama Canal, 1870–1914*. And, he says, you feel you are riding a slow elevator, except that the elevator is your whole ship.) In the second chamber the water rises again, delivering the ship into a third chamber, where the process is repeated a final time. Ultimately, the boat rises in three steps to a height of eighty-seven feet above sea level.

After passing through the Gatun Locks, the ship travels across Gatun Lake and the isthmus until it nears the end of the canal and the Pedro Miguel Lock. Here, and in the next set of locks called Miraflores, the boat is lowered back to sea level as water pours out of the locks. The ship then takes up its voyage on the Pacific side of Central America, in the Gulf of Panama. There is often a backlog of ships waiting to move through the Panama Canal, but once en route, ships only need about eight hours to make the passage from ocean to ocean.

?

How do they make a breed of dog "official" at the American Kennel Club?

AS THE SELF-ORDAINED ARBITER of canine society, the American Kennel Club maintains standards that would be the envy of the Gilded Age's Newport set. Of the four hundred and fifty or so dog breeds that exist in the world today, the AKC deigns to recognize only one hundred and thirty. For a breed to gain official recognition, it must prove that it is genetically chaste; its bloodlines must be untainted by mixed marriages. For individual dogs, admission to the exclusive club depends on two factors. First, Rex must produce pedigree papers that show his bloodlines have been pure for at least three generations; and second, a witness must attest that the hopeful dog conforms to the stringent written standards of his breed. These standards regulate such cosmetic features as eye color, ear shape, overall size, coat thickness, and tail length, but disregard entirely the applicant's ability (or inability) to hunt, swim, or fetch. The American Kennel Club is not a meritocracy.

Breed standards, the ideal against which individual dogs are measured, are written by AKC member clubs devoted exclusively to a certain type of dog. The chow chow club, for instance, decides what a chow chow should look like. By codifying a breed's desirable physical traits, the "parent" clubs make it easy to recognize a "champion" dog: the champion is the one that conforms most exactly to the written standard. Once a dog has been judged a "champion" at a show, it becomes the living aesthetic yardstick by which all other members of its breed are measured, and, consequently, a busy stud or brood bitch as breeders scramble to take a dip in its gene pool.

Breeds are produced by a process of genetic winnowing that limits variability by making matches among dogs that possess similar favorable traits. To get popular "purebreds," many commercial breeders employ a method called "linebreeding," a euphemism for inbreeding: desirable dogs are mated with their desirable grandparents or even parents to produce cosmetically perfect litters. Natural variability is curtailed, but even so, experts admit that the only way to really recognize a "purebred" dog is to examine its pedigree. Hence the necessity of the AKC registry, a sort of canine Mayflower Register.

Rigorous inbreeding creates dogs whose only function is to conform to the highly arbitrary breed standards maintained by the AKC. As a result, some heavily sought-after breeds have been genetically altered. Cocker spaniels, for example, have pretty much exchanged their hunting ability for nice, long coats. Even worse, inbreeding can endorse recessive traits in a dog, leading to a variety of health problems such as skin conditions, organ defects, and mental instability.

The American Kennel Club was formed in 1884 by a group of sport hunters who wanted to organize various existing breed-specific clubs. The original AKC registry listed about five and a half thousand dogs of impeccable pedigree. The idea caught on among the dog-buying and -showing public, and in the 1950s, when more than five million dogs were being registered every year, the AKC adopted strict procedures for recognizing breeds. To be recognized by the AKC, a breed needs first to have been registered domestically for a few decades or, in the case of a foreign breed, for thirty years. Second, there must exist a club devoted to the breed, and the club must have at least one hundred members who own among them at least one hundred of the animals, all of which meet the sponsoring club's physical specifications, as approved by the AKC. Once a breed meets these requirements it must pass through a development period, typically lasting from two to five years, during which it is allowed to enter, but not score, in AKC dog shows. Barring any mishaps or lack of commercial interest, the AKC will make the new breed official—i.e., raise its status from registered to recognized—at the end of the trial period.

Some breeds, such as the Australian kelpie, remain unofficial breeds for years and years, never receiving official status. Usually, however, such a snub is the choice of the sponsoring club, whose members want to cash in on the visibility to be gained from AKC registration, but don't want to engage in the harmful overbreeding necessary to receive official recognition.

?

How do they write headlines at the *New York Post*?

HEADLINES SELL THE *New York Post.* A good one can boost sales by thirty thousand copies in a single day. At forty cents an issue, that's $12,000 in additional revenue for Manhattan's highest-street-sale-circulation tabloid. Even more, a well-received "wood"—an idiom dating back to when headlines were carved from blocks of oak—massages the ego of whoever writes it, which could be one of many reporters and editors who gather around the news desk every weekday evening, or even a copy clerk who happens to walk by and proffer an offhand suggestion. Headline writing at the *Post* is a team sport; old-timers remember each one and the person who wrote it the way veteran baseball announcers recall stats of retired outfielders.

The best way to explain how headlines are written at the *Post* is to take you to the large red-carpeted city room near the southeastern tip of Manhattan where dozens of reporters and editors sit opposite each other at gray metal desks. The day is December 9, 1991. The time is about seven-thirty P.M. Breaking stories include the following: hostage Terry Anderson's return home to New York after six years of captivity in Lebanon; CIA director Robert Gates's warning of pending civil war in the Soviet Union; the conviction of four teens in the fatal subway stabbing of Utah tourist Brian Watkins after the U.S. Open in September 1990; the attempted shakedown of Woody Allen's film crew by a black activist; the crumbling of the beleaguered empire of Robert Maxwell, late owner of the *Post*'s chief competitor *The Daily News*; and finally, William Kennedy Smith's defense testimony in his Palm Beach rape trial.

Succumbing to a temptation they have resisted for the nine days since the well-publicized (and well-covered) trial began, Managing Editor Jimmy Lynch and Executive Editor Lou Colasuonno decide to put Smith on page one. While his testimony was unemotional and smug, it was still credible. Moreover, he has offered stark details of his sexual activity on the night in question. The *Post* can't resist.

The initial layout reads "WILLIAM'S LAST STAND," but editors quickly decide it sounds as if the defendant is losing, which by most accounts is not the case. In jest, Colasuonno offers "WILLIE'S WET DREAM," a tame turn-of-phrase in comparison with his earlier bid, "MISTER SOFTIE." Metro Editor Richard Gooding suggests a film theme: "SPLENDOR IN THE GRASS." A light applause fills the room when Gooding then suggests "WILLIE WHIPS LASCH," but the suggestion is soon scrapped because no one knows if most readers are familiar with the name of the Florida prosecutor trying the case.

The winner: "WILLIAM TELL," an earlier recommendation by Day City Editor David Ng. Pushing his luck, Ng suggests superimposing an arrow through Smith's head—no go.

A kicker below the headline lets readers know what to expect inside the paper. In this case, it is four brief but telling lines:

- Kennedy Grilled 4½ Hrs.
- Says They Had Sex Twice
- Denies That It Was Rape
- Prosecutor Gets Nowhere

According to Colasuonno, *New York Post* headlines are like the windows at Macy's. "If you don't like what you see," he says, "you're not going to come into the store." The *Post*'s subscription base is minuscule. Without a provocative headline, an issue will rot on the stands.

Colasuonno admits that there are some stories that merit appearing on page one, but cannot be said in few enough words. In such an instance, this story is teased on the bottom of the front page in a small box called a tracer, which tells the reader where it is covered inside. "If we can't find the right words for it," he says, "it's just not a wood."

Although the *Post* has been accused of insensitivity to delicate issues, Colasuonno insists that he and other editors try to be compassionate toward human tragedy. "We don't laugh at dead bodies, crime and murder," he says.

But if there is a chance to have some fun with a page one story, Colasuonno and the rest of his jeans-and-turtleneck crew are more than happy to twist the truth for a good laugh. During the week in February 1990 when Donald and Ivana Trump's divorce became public fodder, the *Post* broke sales records with such woods as "GIMME THE PLAZA," and "THEY MET IN CHURCH," the latter an unsubstantiated rumor that The Donald met his then unknown girlfriend in a house of God, which was untrue.

But, as Colasuonno says, sex, along with politics and celebrities, sells papers. And any combination of the three will help the tabloid fly off newsstands during the morning commute to work.

"Sex is sex," he says, "and the temptations are the same for the beer-drinking working man as they are for the most highly paid athlete and celebrity or the most powerful politician. They all put their pants on one leg at a time."

?

How do they find arbitrators for baseball contract arbitrations?

LIKE ANY OTHER BIG BUSINESS, professional baseball is divided between labor and management. Players are labor; they belong to the Major League Baseball Players Association, a trade union. Baseball team owners represent management. The Commissioner of Baseball oversees an organization called Major League Baseball, which acts as the governing body of the professional sport. Basically, a player who has been on the same major league team for three years, or who has been playing in the big leagues for six years, is eligible for arbitration if he cannot reach an agreement with his team when his contract expires.

Generally, the way salary negotiation works in baseball is that the player, through his agent, requests a certain fee, say $1 million a year for three years. The team's general manager then makes a counterproposal, say $750,000 a year for two years, with a signing bonus of $200,000. If the player and the team are unable to reach an agreement, and the player has put in the requisite years of service, the player can elect to arbitrate. This means both he (labor) and the team (management) will be given an opportunity to present their cases to an unbiased third party, the arbitrator. After hearing each side's arguments, the arbitrator must pick the figure he thinks is most fair. He must choose either the figure submitted by the player, or the figure submitted by the team. No compromises are allowed.

Where does one find an arbitrator? The American Arbitration Association, founded in 1926 and based in New York, maintains a roster of eligible, professional arbitrators. Its self-described function is to "provide administrative services for arbitrating, mediating, or negotiating disputes, and impartial administration of elections." The association provides panels of arbitrators for parties involved in contract disputes. Its services are called upon not just for baseball, but for disputes in any business. In fact, baseball contract disputes make up only a small portion of the association's work load.

When the association receives a request for an arbitrator, it submits a list of qualified candidates to the parties involved in the dispute. The list includes comprehensive biographical and professional data for each individual. In a baseball contract dispute, the Players Association and Major League Baseball, acting on behalf of the owner, must jointly select an arbitrator. They start by eliminating from the list of candidates anyone who might have a conflict of interest or who might in any way be prejudiced. Major League Baseball, for example, might object to an arbitrator who has heard disputes in the past and always taken the side of the players. The Players Association might harbor reservations about an arbitrator who has a lot of corporate executives for friends. After whittling down the list, the Players Association and Major League Baseball decide on an arbitrator to hear their cases. Hearings occur during the first two weeks of February, after which the designated arbitrator hands down his decision. The decision is legal and binding; no further appeal is allowed.

?

How do they make margarine taste like butter (almost, anyway)?

"EVERYBODY HAS A DIFFERENT interpretation of butter flavor," says Dr. Peter Freund, manager of product applications at Chr. Hansen's Laboratory, Inc., in Milwaukee. Your interpretation is influenced by the butter you were raised on, the type of dairy cows bred where you grew up, the fodder they ate, and even the amount of time butter was allowed to sit during processing.

While butter is derived from animal fats, margarine—butter's close imitation—is made from the oils of vegetables. In an attempt to minimize the difference in taste, margarine manufacturers buy from laboratories such as Hansen's a host of pastes and liquid flavor additives. (Hansen's offers "theater butter," "buttered corn-on-the-cob," and so on.) These flavor combinations are added in trace amounts to margarine to produce what taste specialists call flavor and aroma "notes." The "top notes" are made of quickly evaporating compounds that work as much on your sense of smell as on your sense of taste. The "bottom notes" are derived from butterfat and free fatty acids. These provide the round, rich taste you associate with butter, a flavor you sense in the back of your throat.

The strongest top note in butter comes from a yellowish-green liquid called diacetyl—one of the products of bacteria fermentation of dairy foods. Butter naturally contains "friendly" bacteria and diacetyl, but margarine doesn't, so flavor companies make up bottles and drums of the liquid in combination with other flavors.

One common method for doing this uses a "starter distillate": first, bacteria are grown in milk and then the diacetyl and other flavors are steam-distilled off in highly concentrated form. Another method uses bacterial fermentation on a nondairy medium; a third synthesizes the diacetyl chemically.

The free fatty acids added to margarine to replicate butter's strong bottom notes often come from real cream or butterfat. But before you health enthusiasts gasp in horror, let it be said that the fat is treated with an enzyme called lipase to break off fatty acids, which in the right combination give the taste you recognize as buttery. The source for the lipase has a lot to do with getting that combination—a favored one is the salivary glands of kid goats.

Finally, salt makes up about one and a half percent of butter, and is usually added in about that percentage to margarine. You'd notice a difference without it, except that extra amounts of diacetyl are commonly added to low-salt and no-salt margarines to jack up their taste even more.

?

How do so many Japanese play golf in a country with so few golf courses?

GOLF HAS BECOME something of a national pastime in Japan, claiming more than 12.5 million enthusiasts. They vie for playing time on the land-starved country's few courses, which offer about as much elbow room as Grand Central Station at rush hour. These Japanese golf courses are unfathomably expensive—membership at Koganei Country Club in Tokyo runs in the neighborhood of $2.5 million, plus annual dues.

On weekends, foursomes begin lining up as early as two in the morning and by dawn, when the first group starts, crowds numbering into the hundreds are not unusual. And these are the lucky men (many courses exclude women) with reservations. Even the most rag-tag courses are booked up to three months in advance.

Japanese business policies further aggravate the weekend congestion problem. Executives are expected to play golf, so much so that they are considered corporate liabilities if they don't. Many companies, however, will not permit employees to play during the week. Not to golf is a grave dishonor, but to have a low handicap is worse. It means you've been neglecting work to practice.

To satisfy their yen for the sport, most Japanese golfers frequent the multitiered driving ranges that have sprung up in densely populated urban centers. State-of-the-art Shiba Golf, in downtown Tokyo, can accommodate 155 people on its three-story range. After golfers hit their shots into a 280-yard rubber field, the loose balls are funneled onto conveyer belts and fed into the basement where they are cleaned and dried before being recirculated by means of a giant pneumatic pump.

At less technologically advanced ranges, usually built on rooftops, golfers drive shot after shot into hanging safety nets, and the ball never actually travels more than twenty feet.

Japanese department stores, recognizing a lucrative trend, have pulled out all stops to cash in on the golf craze. In addition to offering complete lines of accessories and clothing—no self-respecting duffer would be caught dead without the traditional polyester trousers and short-sleeved shirt—some stores now feature indoor practice ranges. At the Sheibu Big Box, on the fifth floor of a Tokyo clothing and sports shop, there is a glass-enclosed sand trap, where real balls are blasted out of real sand and into nets.

Video golf, played live with equipment against a projected backdrop, is another popular simulation of the real thing. Players hit off a tee into a large movie screen on which is projected an image of a world-famous golf course, like St. Andrews or Pebble

Beach. Three cameras record the speed, spin, and trajectory of the ball as it leaves the club, and feed the information into a computer that calculates where the ball would have landed on the actual course. After the image is readjusted, the player again tees off into the screen, repeating the process until he reaches the green, after which he putts the ball into a cup in the floor.

To squeeze in extra practice time during the morning commute, Japanese businessmen have taken to swinging their umbrellas on crowded subway platforms as they wait for their trains, a practice that has resulted in a number of personal injury suits.

Despite the golf craze, called *golf-kichigai* locally, only a small fraction of the enthusiasts ever really play. According to one story making the rounds, during a recent trip to Japan, a professional player stopped in at one of the urban driving ranges and spotted some guy swinging perfectly.

"What's your handicap?" the pro asks him.

"I don't know," the guy says.

"Come again?" asks the pro.

"I've never played on a golf course," the guy says.

How do they set the price on a new public stock offering?

WHEN BILL GATES III WANTED to cash in on all the work and effort he put into Microsoft, the computer software company he founded, he took advantage of something called an initial public offering, or IPO. Today the thirty-five-year-old Harvard dropout is worth $3.9 billion.

What is an IPO and how does it work? An IPO is a corporation's first offering of stock to the public. It is also almost invariably an opportunity for the existing investors and participating venture capitalists to make big profits, since for the first time their shares will be given a market value reflecting expectations for the company's future growth.

When a private company decides to sell shares to the public, it usually hires an investment banking firm, such as Morgan Stanley or Goldman Sachs. This firm assumes the risk of buying the new securities from the corporation at a fixed price and selling them to the public at a markup, the markup representing the firm's profit. The investment banking firm often diffuses the risk among a number of other underwriters called a syndicate.

Determining the share price of a new issue is tricky business. The company and its investment banker often use a Wall Street formula called a price-earnings ratio, or P/E, which is the price of a stock divided by its earnings per share. A company, for instance, whose stock was selling for $20 a share and had annual earnings of $1 a share would have a P/E of 20.

In layman's terms, the P/E indicates how much investors might be willing to pay for a company's earning power. A stock going public with a high P/E means the company going public has good future earnings potential. High P/E stocks—those with P/Es over 20—are typically young, fast-growing companies. Low P/E stocks are quite the opposite and tend to be in mature industries or sectors of the economy that have fallen out of favor with investors.

Setting the IPO price involves a lot of guesswork on the part of the investment banker. If the investment banker believes that investors will perceive the company as a rising star, he will set a price with a high P/E. If he thinks investors will be only lukewarm about the company's future earnings potential, he'll have to consider a lower P/E and thus a lower stock price. If, for example, a hot, young biotechnology firm with high earnings potential is going public, investment bankers would argue that the stock deserves a high P/E. If they expect the stock to earn, say 50 cents a share next year, they might set the share price at $15, or a

P/E of 30. For the investor, that's only about a 3 percent return on each share of stock. But if the company's earnings keep growing quickly in the years to come, that high price may be well worth it.

Private companies like to go public when the stock market is doing well. That way they'll be able to raise more money for each dollar of projected earnings. If, however, the market is in a slump, companies usually wait to go public until a bull market returns.

Once a company decides to go public, it's the underwriters who take most of the risk. That's because the underwriters make the profit on the difference between the price they pay to the issuer for the new stock and the public offering price. When there's strong demand for an IPO, the underwriter will sell all the new stock and make a handsome profit. But if there's weak demand for the stock—perhaps the price was set too high—then the underwriters are stuck with stock worth less than what they paid for it.

?

How do they know Jimmy Hoffa is dead?

AT ABOUT TWO in the afternoon on July 30, 1975, former head of the Teamsters Jimmy Hoffa left his suburban home in Bloomfield Hills, outside Detroit, and drove to a nearby restaurant called Machus Red Fox. Some forty-five minutes later, witnesses say, several gentlemen in a maroon Mercury sedan arrived at the restaurant, picked up the erstwhile union boss, and drove away. Hoffa has not been seen since.

The FBI maintains that Hoffa met his maker that day, and the courts, which in 1982 declared Hoffa "presumed dead," take the FBI at its word. This is somewhat problematic, since the best way to prove death is to produce a corpse, and in Hoffa's case the authorities still have not done that. Lack of corporeal evidence, though, can be explained. In 1981, Charles Allen, a mafia hit man, testified before the Senate that, "Jimmy was ground up in little pieces, shipped to Florida and dumped in a swamp," in which case there would be no body. Another theory, favored by investigators, is that after slaying Hoffa, the men in the maroon sedan delivered his body to a mob-controlled sanitation plant in Hamtramck, Michigan, where it was eliminated. A number of variations on the disposal theme have been forwarded at one time or another. One claimed Hoffa's body was incinerated in Detroit; a second offered that Hoffa was first garroted and then run through a meat processing plant. As recently as 1989, a former organized crime associate announced that Hoffa was interred under the football field at Giants Stadium, in East Rutherford, New Jersey. The fact that no one quite agrees on the whereabouts of the corpse, or the method of disposing of it, suggests not Hoffa's survival but rather the morbid cleverness of mob executioners.

One point on which everyone does agree (except, perhaps, the suspect who swore Hoffa "ran off to Brazil with a go-go dancer") is that Hoffa was murdered. And for that crime, the feds have a motive: they argue that Hoffa planned to knock off Frank Fitzsimmons, the man who replaced him as head of the Teamsters in 1967 when Hoffa was sent to prison for jury tampering. The mob got wind of Hoffa's plot and, since they did not want the former honcho interfering with lucrative new friendships, killed Hoffa before he could get to Fitzsimmons. The prime suspect is Anthony "Tony Pro" Provenzano, a mafioso and ex-Teamster vice-president from New Jersey who was supposed to meet Hoffa at Machus Red Fox the day Jimmy vanished. Tony Pro had never shown up for the meeting. Provenzano, who in 1978 was convicted of ordering the murder of another union official, is now in prison. The FBI also has evidence strongly suggesting foul play: traces of Hoffa's blood and hair were detected in the recovered abduction car. All of this was good enough for the courts, and in 1982, at the petition of Hoffa's son, they declared Hoffa "presumed dead." In Michigan, where the younger Hoffa filed the fateful petition, it takes three years for a presumed dead person to become officially dead, and so in 1985, ten years after he was last seen, Jimmy Hoffa legally died.

?

How do they get rid of gallstones without operating?

GALLSTONES DEVELOP when bile in the gallbladder, a muscular little digestive organ tucked away behind the liver, becomes oversaturated with cholesterol. The gallbladder stores digestive bile from the liver, and too much cholesterol causes the bile to crystallize, a condition that leads to gallstones. Without treatment, gallstones can reach a size of two inches in diameter. Each year, about half a million of the twenty million or so people in the United States who suffer from gallstones require serious medical attention for their attacks. Surgical removal of the gallbladder remains the most common procedure, but in recent years some less painful and less expensive treatments have been developed. Lithotripsy gets rid of gallstones by pulverizing them with shock waves; ether dissolves gallstones when it is injected directly into the gallbladder; a drug called ursodiol can be taken orally to dissolve gallstones.

Lithotripsy, a word whose Greek roots mean "stone grinding," was developed in the 1970s in Germany to treat kidney stones, and has since been adapted for use on gallstones in this country. The treatment works on the same principle as earthquakes: shock waves move by displacing particles in their path; if shock waves displace enough particles when they pass through a solid object, the object will collapse. With gallstones, it generally takes about an hour and as many 1,400 shock waves to crush the stones into grains of sand small enough to pass through the gastrointestinal tract.

Before shock waves can be aimed at a patient's gallstones, the exact position of the stones within the gallbladder must be mapped by ultrasound. Once the doctor has located the offending stones, the patient, usually under mild sedation but not anesthetized, is connected to the lithotripsy machine via a water cushion. The machine generates shock waves, which pass easily through the cushion and the patient's body (human tissue has roughly the same density as water) until they reach the stones. The stones offer more resistance to the shock waves than healthy tissue does; as a result, the gallstones are jarred violently. Eventually, the gallstones collapse. The patient remains fully conscious throughout the procedure and can usually feel a jolting sensation within the abdomen, but no pain. Recovery from lithotripsy is virtually immediate, whereas it may take a patient more than a month to recover from surgery.

Ether treatment, whereby a patient's gallbladder is cleaned by direct injection of liquid ether, requires hospitalization for the duration of the treatment, usually a few days, but is still far less severe than surgery. It is an efficient procedure, akin to saving a dirty cooking pot by scrubbing it with a cleansing agent rather than throwing it away. Ursodiol treatment, like ether, dissolves gallstones; while ether does it almost instantly, ursodiol treatment takes six to twenty-four months, the advantage being the patient need not be hospitalized.

?

How do they make telephones for the deaf?

A TELECOMMUNICATION DEVICE for the deaf, or TDD as they are commonly called, allows a conversation to be typed over the phone lines, permitting people with impaired hearing to read incoming messages. The devices, which resemble portable electric typewriters, are equipped with keyboards, light-emitting diode (LED) character displays that show the text electronically, and modems that convert TDD impulses into acoustic tones. Instead of talking into the phone receiver, a TDD user places the telephone handset in the TDD modem and types out his message. The modem converts the typed message into electric impulses and sends them through the phone line to a receiving TDD, which translates the impulses back into words and displays the message on the receiver's screen. The recipient can then type out a response and send it back to the person on the other end of the line. The setup is akin to a personal Western Union, enabling a deaf person to send and receive telegrams at home.

Like other electronic appliances and computers, TDDs are plugged into ordinary wall outlets; they can be connected to any phone. Most of the newer models are battery operable, a feature that enables them to go and be used anywhere, even at public pay phones, where there might not be a wall socket. The latest portable units are only slightly larger than a calculator, small enough to slip inside a jacket pocket or a purse.

In the United States, the two largest manufacturers of TDDs are Krown Research and Ultratec, each of which produces a wide variety of machines. Depending on the model, a TDD might include such functions as a built-in printer for producing hard copy of conversations, and "direct coupling"—linkage through a standard modular telephone cord—which eliminates the need for a traditional telephone set. Some of the more advanced devices can be hooked up to computers or television sets to display more lines of type.

A phone system, no matter how advanced, is useless if it doesn't inform the owner of incoming calls. To get around the obvious problem of how to let a deaf person know when the phone is ringing, TDDs are commonly used in tandem with signaling devices that cause lights to flash when a call comes in. The signaler is a small free-standing box, about the size of a transistor radio, into which a lamp can be plugged. When the phone rings the lamp begins to flash. The main signaler, which must be plugged in near the phone, can also transmit messages to receivers arranged throughout the house, setting off multiple blinking lamps.

The biggest restraint to TDD use is that the devices can communicate only with other TDDs, a considerable dilemma for the nation's twenty-one million deaf and hard-of-hearing people. Through the efforts of such groups as the National Association for the Deaf, though, more and more businesses and public facilities, like airports, are installing TDDs. The devices range in price from about $200, for a basic unit, to $650, for a top-of-the-line model.

?

How do banks make money off of credit card purchases even when you don't pay interest?

SINCE MOST PEOPLE DON'T pay their credit card bills in full each month, the bank that issued the card usually *can* charge interest, and that money, called finance charges, constitutes by far the greatest percentage of income for the bank. But let's say you are conservative with your credit, don't run amok by overspending, and don't let the bills slide. You won't provide the bank with any interest on your tally of purchases, but it is assured of other income nonetheless.

The next two biggest sources of income for banks that issue credit cards are the annual fee and the merchant discount rate. The bank may charge you anywhere up to about $50 per year just for the privilege of having the card. However, in a society glutted with plastic, some banks are trying to lure customers by literally giving their cards away, with no annual fee. American Express, on the other hand, which is a company and not a bank, continues to charge a "membership fee" of $55 a year. It must charge a higher rate because the American Express card is a charge card rather than a credit card: you must pay your entire bill each month, so the company receives no income from finance charges.

In order to draw the largest possible number of consumers, most industries and services today, from restaurants to furniture stores, allow you to pay by credit card. The merchant himself gives up a small percentage of his income for this service. Here's how it works. A merchant who wishes to let his customers pay with Visa or MasterCard first makes an arrangement with a merchant bank. This bank in turn works with a card-issuing bank, say, Chase. The two banks may in fact be one and the same, but the functions are different and independent of each other. The merchant bank decides, within a fairly narrow range, what percentage of the face value of the purchase will be retained by the card-issuing bank. It may also lease the credit-card verification equipment to the merchant, especially if his business is small and he can't afford to buy it outright. Exactly what the merchant's "discount rate" is (i.e., what percentage he pays), depends on a number of factors, such as type and volume of industry. Obviously, a large chain such as Macy's will pay a lower percentage than an independent merchant with a single store. The type of data transmission facilities also have a bearing: if the merchant reconciles his books and accumulates his drafts only every few days and submits them by mail, he will pay a higher rate—the "paper rate"—than if he submits them electronically by computer, which performs the function daily, even instantly. (One reason banks prefer the faster service is it helps prevent fraud.) Typically, the merchant discount rate for MasterCard or Visa is 1.8 percent.

Now let's say you buy a sweater from a small boutique for $100.00 with your Chase Visa card. The owner of the store submits a draft to the merchant bank for $100; it keeps a very small amount—about 35 cents in this case—as a fee for processing the draft and delivers the transaction to the card-issuing bank. Chase retains the balance of that 1.8 percent—$1.45 out of the $100.00. This fee is intended to cover Chase's costs of posting and processing your account, billing you, and so on. Chase is not making a profit, though. "It's a break-even scenario at best," says David Anglin of Visa U.S.A. And MasterCard's president Alex Hart admits, "The fees we charge merchants don't even cover our costs."

Not so at American Express. That company charges merchants a whopping 3.98 to 4.5 percent, which comfortably covers the company's costs and then some. The significantly higher rate is why some merchants don't subscribe.

Finally, card-issuing banks accrue some income from "service fees." If you exceed your credit limit, the computer sends out a red alert and the bank charges you some five or ten dollars. If you bounce a check to the bank or fail to pay the minimum amount on time, again you'll be charged. The amount of these nuisance fees—so called by consumers—bears some relation to how much it costs the bank to handle the problems. The big profits still lie in the finance charges.

?

How do they choose cartoons for *The New Yorker*?

COSMOPOLITAN CARTOONS have long been a trademark of *The New Yorker*. You can find the cartoonists' names listed modestly at the end of the magazine's table of contents under the heading "Drawings." The term is a holdover from the late 1920s and '30s, when *New Yorker* cartoons were sometimes written by one person and drawn by another. Now, of course, this is no longer the case; each drawing is created by a single artist. And art editor Lee Lorenz, himself a contributor, comfortably refers to them as "cartoons."

The New Yorker has a group of about sixty-five artists who contribute regularly to the magazine. Each artist submits twelve to thirty cartoons every week to Lorenz, who narrows the field and shows his choices to the magazine's editor. Together they settle on the best, selecting some for immediate use and others to be published later. Each weekly issue contains about two dozen cartoons. The choices for future use are stored in a reserve bank, filed under seasonal themes or categories such as "heaven and hell," "married couples," "cocktail parties," "olden times," "talking animals," and so forth.

Some of the cartoons in the magazine—albeit a very small percentage—are created by artists outside the regular group. Lorenz himself reviews all unsolicited submissions, which pour in at a rate of hundreds and even thousands a week. He's on the lookout for new talent, in the form of "a recognizable, individual graphic style and personal point of view." If he likes the work in a new portfolio, he meets with the artist and starts buying one cartoon at a time. Eventually he may invite the artist to sign on with the magazine—but not very often. He has offered only eight new contracts for cartoonists in three years, and the established artists tend to stay on. William Steig and George Price, for example, have been keeping *New Yorker* readers in stitches since the late 1920s.

"People at *The New Yorker* tend to dismiss the suggestion that there is such a thing as a *New Yorker* style," wrote Louis Menand in *The New Republic* magazine in February 1990. "The requirement for a story, they say, is that it be well written; the requirement for a drawing is that it be funny."

Lorenz would agree. Asked to describe the humor in *New Yorker* cartoons, he replies, "I can't. . . . It's too subjective." He disagrees with the term "highbrow"—although he agrees the magazine is aimed at the well-educated, well-heeled, upper-middle-class reader. The humor ranges, he says, and "the audience [for cartoons] extends beyond people who love *The New Yorker*. People can go to the magazine and get a laugh even if they're not interested in fact, fiction or reportage."

New Yorker cartoons do in fact range more widely than ever before. "The world has changed a lot," Lorenz says. "You can't outrun your audience, but you certainly have to keep up." To that end, he has introduced cartoonists with new styles. Roz Chast, for example, contributes some wonderful cartoons with a very personal, postcard-like sensibility. And in his cartoons, Danny Shanahan makes hip verbal and visual puns: "Chicken à la King," the caption reads beneath a picture of a chicken dressed as Elvis Presley.

?

How do they make rechargeable batteries?

ALL BATTERIES WORK by converting chemical energy into electric energy. The difference between primary, or nonrechargeable, and secondary, or rechargeable, batteries lies in the specific chemical reactions each battery type uses to produce an electron flow, which in batteries is electricity. In primary batteries the chemical reactions cannot be reversed, thus a primary battery can be discharged only once; the chemical reactions that give a secondary battery its power are reversible, allowing electricity to be regenerated after each discharge.

A battery cell is composed of an anode, which is a negative electrode, and a cathode, a positive electrode. The two electrodes are separated by an electrolyte, a chemical medium through which electrons can travel from the anode to the cathode. The cathode is usually made of a metal that naturally releases oxygen ions, while the anode is usually made of a type of metal capable of oxidizing, or mixing with oxygen. The oxygen released by the cathode is absorbed by the anode; such oxidization in the anode generates excess electrons, which produces a chain reaction: the flow of excess electrons from the anode to the oxygen-losing cathode. The flow of excess electrons from anode to cathode triggers the cathode to release more oxygen, which in turn causes the

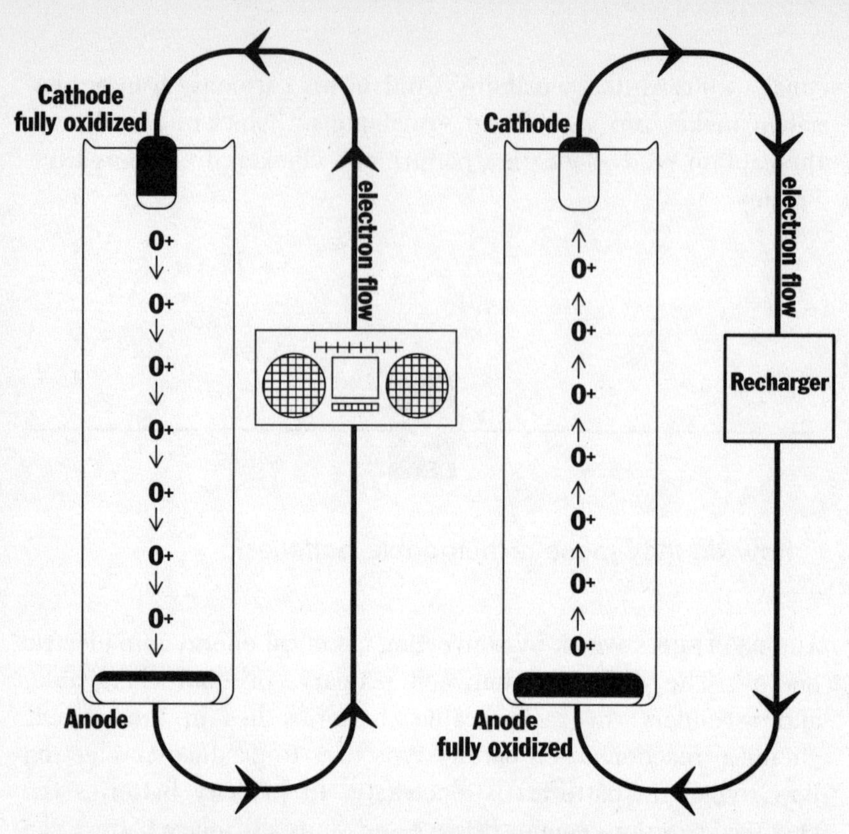

The chemical reactions that power rechargeable betteries are reversible. As the battery discharges* (left)*, the cathode releases oxygen to the anode, causing it to generate electricity. Once the anode is fully oxidized, the battery must be recharged. In recharging* (right)*, the anode releases oxygen back to the cathode.

anode to oxidize, thereby generating more electrons. Think of a battery as a microcosm of a self-contained university: the cathode is like an undergraduate college which continuously graduates students (oxygen) who enroll in graduate school (the anode) where they are transformed into professors (electrons), who stream over to the undergraduate college (the cathode) to find jobs.

In a primary battery, the migration of electrons can only occur in one direction, from anode to cathode. When the anode has absorbed all the available oxygen, it stops producing electrons, and the flow of electricity ceases. In a secondary battery, once the fully oxidized anode has stopped producing electrons, the chemical reactions that produced the electron flow can be reversed, thereby returning the anode to a deoxidized state in which it can once again absorb oxygen and produce electricity.

A nonrechargeable battery is sort of like a really inexpensive, disposable videocassette (if there were such a thing) that cannot be rewound: after such a cassette had been played there would be no way to get back to the beginning. A rechargeable battery discharges and recharges just as a standard cassette plays and rewinds.

The two most popular types of rechargeable batteries are lead-acid and nickel-cadmium. Lead-acid batteries, of which the typical car battery is the most common example, have lead (Pb) anodes and lead dioxide (PbO_2) cathodes. Nickel-cadmium batteries have nickel (Ni) anodes and cadmium (Cd) cathodes, and come in the full range of standard sizes—triple A to nine volt—for use in all the familiar battery-operated devices. Alkaline batteries, the standard throwaway kind, have zinc (Zn) anodes and manganese dioxide (MnO_2) cathodes.

A battery's voltage is determined by the materials used for its anodes and cathodes. Zinc and manganese dioxide combine to produce a voltage of 1.5 volts, slightly higher than 1.2 volts achieved by nickel and cadmium in the standard rechargeable battery. Voltage, however, is reduced with use, and whereas an alkaline starts at 1.5 volts but steadily loses power, a nickel-cadmium battery maintains its 1.2 volts through 90 percent of its discharge. This means rechargeable batteries, although they run down faster than alkalines, provide consistently more power while in operation.

A single charge of an AA nickel-cadmium battery lasts only 25 to 50 percent as long as the total discharge of an AA alkaline battery, which means that a nickel-cadmium battery might go through four cycles before matching a throwaway's life span. But since a nickel-cadmium battery can be recharged as many as a thousand times, its total longevity equals about 250 generations of alkalines.

?

How do they set the minimum bid on a painting that is up for auction?

BEFORE PUTTING A PAINTING on the auction block, the auction house responsible for selling it and the consigner (the owner of the painting) decide upon a reserve price. A reserve price is the price below which the owner is unwilling to part with his painting. It is frequently set at about 60 percent of the low estimated value of the painting. For example, if the estimated value of a painting is a minimum of $1,000,000, the reserve price might be $600,000. There is no universal formula for establishing the reserve, and sellers enjoy some flexibility in setting their bottom line. Legally, the reserve cannot exceed the low estimated value of the painting being sold.

To come up with the reserve price for a painting, an auction firm must first arrive at an estimate of the painting's worth. In the parlance of the trade, an estimate is a determination of the amount a painting is likely to fetch at auction. An estimate differs from an appraisal, which is a written document used for insurance and tax purposes.

A number of factors, including the quality, rarity, condition, and provenance (history of ownership) of the painting, influence the final estimate. Market conditions also come to bear: the art market, like the stock market, tends to be either bearish or bullish. Nineteen eighty-seven, for instance, was a bullish year—one that saw the van Gogh still life *Irises* sell for a record $72 million. Since 1989 the market has quieted down. It's possible that if

Irises was auctioned off today, it would fetch less than what it did in the gaga market of 1987. Another important factor in determining the worth of a painting is historical precedent, that is, the prices attained during the last six months to a year in sales of similar paintings.

Let's say you are a businessman and art collector. You've taken a bath on some ill-advised real estate investments and need to raise cash to cover your debts. You decide to "de-acquisition" (art-world doublespeak for sell) an Old Master painting. The first thing you would do is contact one of the big auction firms like Sotheby's and tell them you want to sell, say, a Titian. Because you are famous and a painting by this sixteenth-century Italian painter is significant, the auction firm will probably send a staff appraiser to your house rather than making you ship the painting to them, which is how it usually works.

The appraiser will examine the painting for integral quality: how it stacks up against other Old Master works and, especially, against other paintings by Titian. Is it a good painting by Titian or not one of his best? Major work or relatively minor? In terms of rarity, any painting by Titian is automatically first-class, because few of his works are in circulation. Then the appraiser will examine the condition of the painting. A work of art that has been damaged in any way is not worth as much as one in pristine condition. If the painting is impaired, the appraiser will have to decide whether the damage is reversible. A slight tear in the canvas, for instance, can probably be fixed; damage caused by sloppy restoration efforts, on the other hand, is permanent and will devalue the painting.

After examining the painting, the appraiser will usually come up with a preliminary estimate. Before the final, pre-auction estimate is made, however, he'll check the going rate for Titians. Probably no Titians will have come on the market in the last six months to a year, so market analysts will look for other Old Master paintings of comparable quality and find out what they sold for and when. The subject matter of the painting will also influence the estimate: the market is always better for top-quality mythological love scenes than for gory crucifixions, two favored Renaissance subjects.

Once you receive an estimate for your painting, you will set the reserve price. The reserve is between you and the auction house; it is not released to the buying public. A number of factors, such as how much money you need, how long you are willing to wait to sell the painting, and the number of bidders expected to turn out for the auction, will influence where you set the reserve. You can set it at any price up to and including the low estimate, but you cannot legally set it higher than the low estimate. This means that if your painting has been estimated at between $20 million and $25 million, the reserve can go up to $20 million. If at the auction bidding fails to reach the reserve price, the painting will be removed from the block and either held by the auction house to be re-auctioned at a later date, or returned immediately to you, depending on what you and the auction house decide to do.

How do antilock brakes work?

AT THE MERCEDES TEST TRACK in Stuttgart, Germany, workers drive these elegant autos at high speeds through slalom courses. Going ninety miles an hour, the driver slams on the brakes and nimbly maneuvers the car through a series of orange pylons. Why don't these pricy vehicles flip off course like so many Matchbox cars in the hands of a four-year-old? Because of the quick response of antilock brakes.

Today it is not only luxury models that can give you the addi-

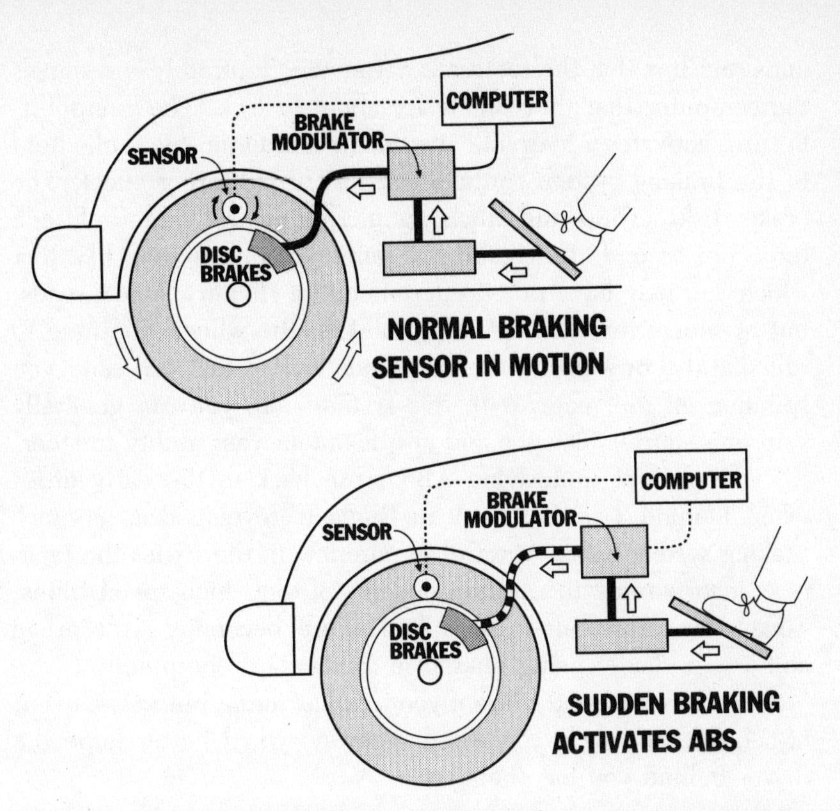

During normal braking, the sensor on the wheel moves smoothly and monitors how fast the wheels are turning. When you slam on the brakes for a sudden stop, the sensor informs the computer, which in turn signals the brake modulator to initiate a rapid braking action—the antilock brake system.

tional safety of antilock brakes. General Motors, for instance, includes them as standard equipment on many of its Pontiacs, Cadillacs, Buicks, and Chevrolets and offers them as optional equipment on other models.

The heart of the antilock braking system (ABS) is a computer inside the vehicle. A sensor located on the inside of the wheel monitors how fast the wheels are turning and feeds this information continuously to the computer. The sensor has calibrations, or small teeth, that run smoothly when the wheels of the car are turning properly. If you brake gradually, your brakes will perform on their own as usual. But if you suddenly jam on the brakes—regardless of how fast you may be going—the sensor in milliseconds registers that the teeth are not moving uniformly and signals the computer that the wheels are about to lock. The computer, in turn, activates a hydraulic modulator that forces hydraulic fluid to the braking system, initiating a sort of stuttering effect. The brake disks clamp and unclamp incredibly fast—up to fifteen times per second. The result is a rapid pumping action, like that which you may have practiced yourself on slippery snowy roads, but of course much faster. The point is, the wheels continue to roll slightly, preventing the hazardous lockup that can send you spiraling off the road. With this system, too, you can generally stop in a shorter distance and also maintain your ability to steer.

The concept underlying ABS arose back in the early fifties when Dunlop Tire Company in England developed an antilock braking system called Maxaret for aircraft. In the sixties the Japanese incorporated the system in some of their high-speed trains. Today, in automobiles, these brakes are becoming widespread and are gradually being phased in as standard equipment.

Any drawbacks? It's fine if your car has them, but in the event of a quick stop on a congested highway, you'd better hope the driver behind you has them too.

?

How do they know television viewers don't cheat during the Nielsen ratings?

NIELSEN MEDIA RESEARCH, the company responsible for compiling information on America's television viewing habits, employs a device called the Nielsen People Meter to get its figures. The electronic meter, which is about the size and shape of a small cigar box, is attached to each of a sample family's television sets, and whenever a set is turned on, the meter automatically records the time and the channel. The meter also documents channel switches and the time at which the set is turned off. Since the meter communicates directly with the television set, going over the head of the fallible viewer, the possibility for cheating is all but eliminated. The machine remembers what people watched until Nielsen technicians collect the data, and if someone was to detach the machine from the television set, it would remember that, too.

Nielsen maintains a field of four thousand sample viewing households, a demographically representative sample of the entire country, and applies the information it receives from them to the country as a whole. If 15 percent of Nielsen sample viewers tune in to a program, the company assumes that 15 percent of the national television audience tuned in along with them; the program would then be assigned a rating of 15. Since 92.1 million U.S. households have televisions, a rating of 15 translates into 13.8 million households. The math works like this: 15 percent (rating) × 92.1 (U.S. households with television sets) = 13.8 million (number of households tuned in).

In order to obtain viewer profiles, the Nielsen People Meter is equipped with a number of personal viewer buttons, one for each sample household member. When a sample viewer turns on the TV set he is supposed to identify himself to the meter by punching his button. Button number one, for example, would belong to Bob, a thirty-five-year-old father, while button number two would belong to Shirley, Bob's thirty-year-old wife, and button number three would belong to their five-year-old daughter, Susie. The viewer identification buttons do allow for the possibility of deception, as there is nothing to prevent little Susie from punching Bob's button every time she turns on *Sesame Street*. Even so, while such deception might throw off the demographic breakdown slightly, it would not affect a show's rating.

Another source of error arises when someone turns on the set and then goes into another room, say, to prepare dinner; since the show is on, it gains rating points even though no one is watching it. If a sample viewer is really intent on disrupting the ratings, he could secretly buy an extra TV for private use, all the while keeping the metered set tuned in to talk shows and late-night movies. Admittedly, such behavior, if it exists at all, is extreme, and Nielsen doesn't waste time worrying about infiltration by the odd crank.

Nielsen collects the information stored in its People Meters nightly by means of a direct phone line to each box. The company usually does this around two o'clock in the morning when everyone is asleep.

The four thousand sample households, selected randomly from a pool maintained by Nielsen, remain pretty much constant. Changes occur when the members move and leave the meter behind, or when a household tires of participating in television research and asks Nielsen to come remove the meters.

?

How do they manufacture genes?

MAN CANNOT MAKE A GENE. He can only emulate one. In fact, scientists are less interested in reproducing actual genes than they are in copying the stuff genes are made of—clumps of DNA that are the blueprints for manufacturing proteins and chemicals imperative for life.

Formulated in 1985, the polymerase chain reaction (PCR) has enabled scientists to make millions of copies of a single strand of DNA in a matter of hours. The technology both hastens and consolidates the more onerous tasks of genetic mapping and se-

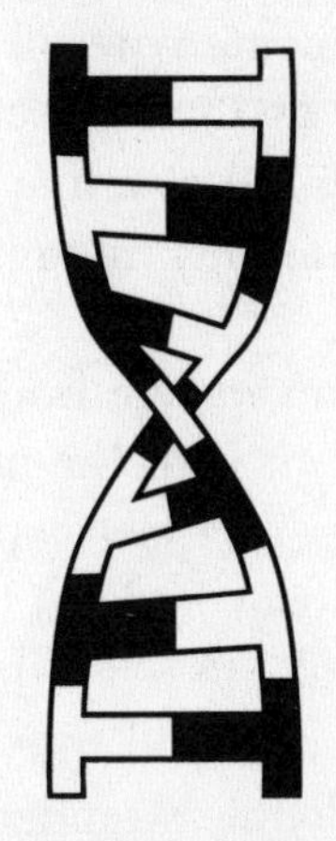

1. ORIGINAL DNA FRAGMENT.

2. HEATING "UNZIPS" DNA INTO TWO STRANDS.

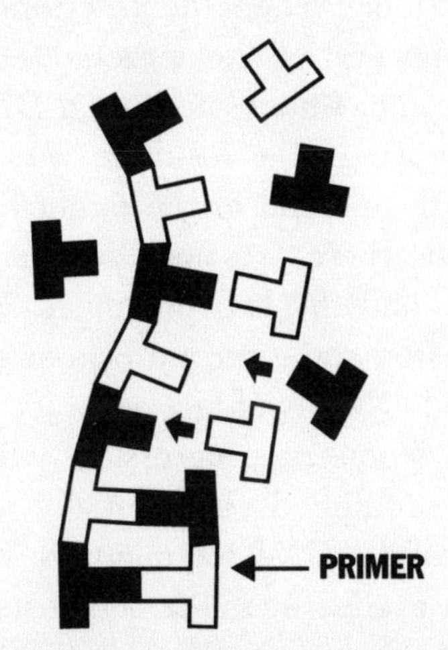

3. PRIMER ATTACHES TO ONE END OF EACH DNA STRAND AND BEGINS BUILDING NEW COPIES....

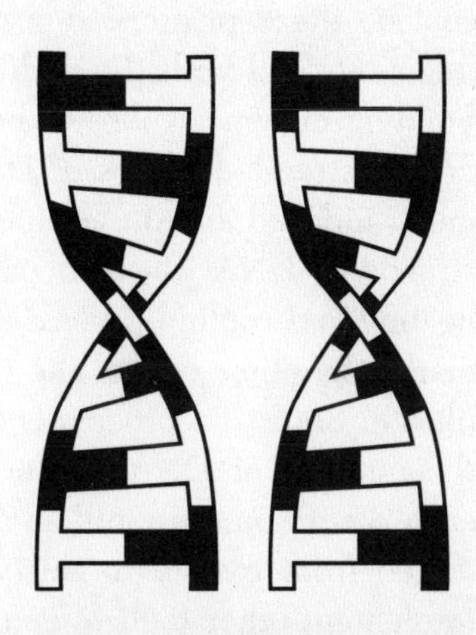

4. RESULTING IN TWO NEW DNA FRAGMENTS READY FOR NEXT HEATING CYLCLE.

Scientists can alter the blueprint of a DNA fragment by this process called polymerase chain reaction, in which a new DNA fragment is introduced and two new DNA fragments are created.

quencing that once could take several years to complete. According to an article in *U.S. News and World Report,* PCR has been used to find genetic defects that cause some types of cancer, to design diagnostic tests for such illnesses as Lyme disease and AIDS, to identify criminals through samples left at crime scenes, and even to recover DNA from frozen mammoths, mummies and other long-dead organisms.

PCR works simply and efficiently. It takes its cue from a process already perfected by every living system: with the aid of an array of enzymes, a cell divides and makes an exact copy of itself—a phenomenon that occurs in the human body millions of times every second. By heating a strand of DNA inside a test tube, a technician separates the DNA molecule and thus creates two complementary strands of the original double-stranded DNA. Next, short, synthetic DNA molecules known as primers are added to the test tube, at which point the process goes beyond the natural function of a living system. When the test tube is cooled, these primers anneal to one end of the targeted DNA fragment, and with the addition of nucleotides, instruct a DNA-building enzyme where to begin copying. The result is two new double-stranded pieces of DNA with one end of the target fragment indexed by the primer. Thus there are now four target strands of DNA: the two parentals and the two progenies. The heating and cooling process is repeated and a second primer earmarks the other end of the targeted DNA, yielding an exact duplicate. Since the process is one of geometric progression, where the number of DNA produced is multiplied exponentially with each repetition, one million copies of the DNA fragment are generated from twenty to thirty such cycles, enabling scientists to implement other tools of genetic engineering to research the specific makeup of the gene.

Transgenic manipulation is another method of producing new DNA. Using a microscope and microsyringe, foreign DNA is injected into animal egg cells so that it can be incorporated with human genes coded to produce either enzymes or proteins and so that its genetic performance can be studied. In what some analysts believe will grow into a $1.5 billion industry, transgenics are expected to help produce leaner livestock and to cultivate animals into "living factories" that manufacture protein-based drugs.

Surely the most ambitious and hopeful sign for the future of genetics is gene therapy, a process that permits "healthy" genes to be injected into the body to combat faulty ones. According to Dr. Michael Blaese, a researcher at the National Institutes of Health (NIH) who is conducting the first human trials in gene therapy, by the middle of the twenty-first century the technology will "be used for treating everything from cardiovascular disease to senility to cancer as well as for many of the more than four thousand recognized inherited diseases, many of which have no effective treatment at all."

Gene therapy brings scientists into the nucleus of the cell, where they can manipulate genetic material and explore methods of treating disease at its source. The ultimate goal of gene therapy is to treat inherited metabolic diseases prophylactically—in other words, to treat genetic deficiencies while a child is in utero so that chances of developing ailments such as cancer and heart disease are completely wiped out. This would be accomplished by repairing a defective gene or adding one which would correct instructions delivered to the cells. Though still in an experimental phase, gene therapy is used to treat such diseases while they are in progress. Some experts believe that in utero gene therapy will be as common as amniocentesis, and instead of just learning about what deformities a child will be born with, or will be predisposed to, the disorders will be corrected before an infant takes its first breath.

As it exists today, gene therapy permits genetically altered DNA to be attached to a disarmed virus which serves as delivery truck to targeted cells. The NIH recently received government approval for what might result in the world's first cancer vaccine. The vaccine will be made from the patient's own genetically altered cancerous DNA, which is then reinjected into the body. This form of gene therapy has been proposed for the treatment of advanced skin cancer, gastrointestinal cancer and kidney cancer.

?

How do they stop you from getting a credit card under a false name?

CREDIT CARD FRAUD WRENCHES millions of dollars every year from the credit industry. And even though that's only a fraction of a percent of credit card issuers' profits, most issuers, such as banks, are careful to review applications. Hypothetically you could call yourself Spot, so long as you paid your bills. But in reality, it's illegal to take a phony name, and you'd get caught: your bank wants to know the real you. The real you is a lot more likely to pay.

Credit card cheats can be pretty creative. Their scams using false names vary considerably. Sometimes they read the obituaries and apply for credit using the name of a dead person. And some actually *have* tried using their dogs' names or made-up names. But the most common ploy by far is to steal somebody's wallet and credit cards, then apply for credit under the victim's name. This works nicely because it gives the thief access to his victim's driver's license, social security card and possibly even such things as a job ID card. The thief fills out an application for credit under the victim's name. He uses the victim's address as "previous address" and uses his own address as "current address." That way, of course, the new credit card gets mailed straight to the thief at his address, which is usually a post office box he can change in a pinch.

Criminal groups tend to apply for credit many times a year using this very method, with all sorts of assumed names. A high number of applications coming from the same address is one clue to a credit card company that something is amiss. In the early days of credit cards decades ago, at least one man collected thousands of cards under false names before he was caught.

So how *do* banks and credit card companies stop you from getting a card under a false name? Using computers and telephones, they monitor your application patterns and catch your slipups. Visa and MasterCard—the two biggest credit card companies in the world—have a joint "Issuers' Clearinghouse Service," which keeps millions of computer records of addresses, phone numbers, number of applications per year, Social Security numbers of deceased people (numbers supplied by the U.S. government) and past fraudulent applications. When you apply to a bank for Visa or MasterCard, the bank is required to run your application through the ICS computer. In a few minutes, your bank knows whether to suspect you of fraud.

When the computer notices, for example, that more than five credit applications have come from your current address in the past year, a flag goes up. The ICS tells your bank, and any other banks that have issued you credit recently, to review their relationships with you. Have you been paying your bills? Are you the person you say you are? Or perhaps the computer notices that you've used the name and Social Security number of a deceased person. You'll have to explain why (and for that matter, *how*) you could still want a credit card when you're dead.

While computer records can catch schemes that follow patterns like these, the one-time fraud can also be caught with a few telephone calls. The thief who stole the wallet is a good example. A bank employee would check the phone book and call the applicant's previous-address phone number. If the person whose name is on the application still lives at the old address, then whoever lives at the "current address" is highly suspect.

Similarly, a call to the applicant's employer verifies that the person actually exists; talking with the applicant where he works can confirm that the application form was filled out by him. Some chiselers set up elaborate answering machine systems, or ask someone to verify their employment. But banks can sometimes get around that tactic by using their own phone books. They look up the main number of the company and call that number instead of the one given on the application.

Still, even massive computer records and intensive telephone inquiries can fail. It's nearly impossible to keep computer files completely up to date; obituaries run before death certificates are filed, for example, and a fraud can take advantage of the lag time. Perhaps the most difficult cheat to stop is one who uses any one of these schemes to apply for credit under a false name, and then maintains good payment patterns for a while before vanishing into thin air.

?

How do members of Congress make themselves sound articulate in the *Congressional Record*?

WHENEVER THE U.S. HOUSE OF REPRESENTATIVES and the U.S. Senate are in session, their public proceedings are published in the daily *Congressional Record.* The *Record* was first published in 1873, and it continues to be the official account of congressional debate.

Each house of the legislature maintains a staff of court reporters, who transcribe the speeches made from the floor by members of the House and Senate. These transcriptions make up the *Congressional Record.* Within half an hour of finishing a speech, a member of Congress is presented with a typescript of it. He then has an opportunity to make sure his words have been taken down accurately before they are included in the *Record.* Members are not supposed to make any revisions that change the meaning of their speeches. The rules of the Congress restrict them to correcting grammar, spelling, and so forth.

When a representative or senator delivers a speech prepared ahead of time, he simply presents a copy of it to the transcribers. Often he won't even bother to check the typescript, since he assumes the transcriber has copied the prepared document accurately. Most speeches, in fact, are prepared in advance, so the speaker will have had all the time in the world to make himself sound articulate.

Remarks made extemporaneously, on the other hand, do not always come off smoothly. When speaking off the cuff, most people, including most legislators, abandon precise, textbook English. They use run-on sentences, mix plural subjects with singular verbs, dangle participles, and commit many other linguistic transgressions. These errors don't necessarily prevent the listener from understanding the point of the speech, but they would seem pretty awful if put into print. When a speaker is ad-libbing, the court reporters record every grammatical foible, including all the "ums" and "ahs." Later, it is the job of *Congressional Record* copy editors to put it all into proper English. The copy editors delete pause words, add words to complete broken sentences, patch sloppy transitions, and introduce punctuation where necessary. Copy editors are not in any way allowed to change the meaning of a speech; they merely clean up the grammar to ensure that the remarks are coherent enough for the printed page.

Occasionally, a legislator will, in the heat of debate, say something universally offensive. When this happens, one of the unruly speaker's colleagues will usually raise an objection and the body will vote by unanimous consent to have the offending words "taken down," or expunged from the *Congressional Record.* Members can also petition to take down their own words. Sometimes, after expressing an opinion in a speech, a member will think better of the opinion and begin to rue his words. He can then request that those words be deleted from the official transcript. Barring any objections from his colleagues—and usually none is raised—the request will be granted. A good example of such after-the-fact tinkering occurred during the Clarence Thomas confirmation hearings. Senator Orrin Hatch, a Republican from Utah and a Thomas supporter, took the floor to defend the nominee. His defense took the form of a direct attack on his avowed friend, Senator Edward Kennedy, who opposed confirmation. Senator Hatch said something like, "And if you believe that, I've got a bridge to sell you in Massachusetts," referring none too subtly to the infamous bridge at Chappaquiddick Island. Hatch meant to imply that Kennedy's judgment was so impaired no one should listen to anything he had to say about Judge Thomas. Later, after he had cooled down a bit, Hatch asked to have "Massachusetts" changed to "Brooklyn," thus transforming his personal and malicious statement into the generic "And if you believe that, I've got a bridge to sell you in Brooklyn." No one objected.

After material has been printed in the daily *Congressional Record,* members of congressional staffs examine it to make sure the words of their bosses have been presented accurately. If a staff member discovers a mistake, he can ask that a correction be printed in the next day's *Record.* The permanent *Record,* the hardbound, comprehensive tome that is published at the end of each congressional session, will then reflect the correction. As is the case with the daily *Record,* senators, representatives, and their staffs are not allowed to revise meanings for the permanent *Record,* only to repair errors in transcription.

?

How do they "make" veal?

ACCORDING TO THE United States Department of Agriculture, a vealer is "an immature bovine animal not over three months of age: the same animal after three months, and after having subsisted for a period of time on feeds other than milk, is classified as [a] calf." Veal meat, then, comes from a younger animal than does beef. Unlike beef, which comes from a grain-fed cow or steer, veal comes from an animal raised for slaughter on a strict diet of milk, milk by-products, or milk substitutes.

Because of the young animal's severely restricted diet, the meat of a vealer is noticeably different from mature beef. While healthy beef is pink, veal, because of the absence of iron from the vealer's diet, has a pale, grayish appearance. Beef is marbled and trimmed with fat; veal comes from an animal that has not yet developed substantial fat, and the flesh retains an even, creamy texture. The soft, slender, and reddish bones of vealers also distinguish them from their older counterparts, whose bones are heavier, whiter, and harder.

The two types of meat are taken from significantly different breeds of cattle. Veal, which is sort of a dairy spin-off, comes from the young of dairy cows, predominantly Holsteins and Guernseys in this country, while beef comes from the larger, meatier breeds, such as Hereford and Angus. Only cows that have recently calved produce milk, a fact that necessitates a high birthrate among dairy cattle. The high birthrate leads to surplus calves, and surplus calves lead to veal cutlets, which, coming from animals not bred for meat production, are more delicate than cuts from big, mature beef cattle.

In recent years, veal has become something of a political issue, with various animal rights activists attacking as inhumane its method of production. They find offensive the fact that veal breeders often pen the vealers in such a way as to prevent their muscles from developing, so that the meat of the animal remains soft and tender. The combination of enforced immobility, iron-free diet, and the young age at which the vealers are slaughtered, is more than enough to turn some people off Wiener schnitzel for life.

?

How do they determine the profit of a Hollywood movie?

IN THE MOVIE BUSINESS, as in every other industry, profit equals gross revenue minus costs. For movies, the biggest money maker is box office ticket sales. Additional income is derived from video sales and rentals. Other movie markets include cable and free TV, airlines, and shiplines, but in general these ancillary outlets provide only a fraction of the income generated at the box office.

The average Hollywood movie costs about $25 million to make, a figure which includes salaries for high-priced stars and directors as well as the blue-collar army of gaffers and bestboys and so forth who work behind the scenes. Add to that $25 million the amount spent on advertising and distributing the movie and you have its total cost. Deduct that cost from the revenue generated at the box office and through sales and licensing fees for video, television, and the ancillary markets, and what remains is profit.

Of the six bucks or so that moviegoers shell out for a ticket to the latest release, only about half goes back to the movie studio. The rest is kept by the cinema. Movie theaters contract with movie distributors for the rights to show a particular movie; generally the distributor collects between 42 and 52 percent of a movie's box office receipts, more or less as a rental fee. Therefore, when the entertainment media reports that some blockbuster made a whopping $40 million at the box office over the weekend, it means that the movie has earned roughly $20 million for its makers. Whether or not that amount is profit, depends on how much the movie cost to make and how long it has been running.

Officials at Entertainment Data, Inc., the independent company that contracts with many studios to compile and release box office figures, report that they get their information straight from the movie theaters. Every night, EDI's large reportorial staff calls the theaters to get figures for those movies that the company has contracted to track. At the end of the week EDI adds up all the numbers and announces them to the parent studios. Then it delivers the news to the press. Some studios bypass EDI and do their own tracking, a pretty big chore considering that at any one time a major movie will be showing on anywhere from 800 to 1,800 screens nationwide.

?

How do they decide who gets nominated for Oscars?

THE ACADEMY OF Motion Picture Arts and Sciences has 4,830 members, each with expertise in some branch of the film industry, who nominate their peers for the various Oscars. Cinematographers nominate cinematographers; directors nominate directors; actors nominate actors; and everyone nominates movies for the Best Picture Award.

The nomination process begins at the end of the Academy year, December 31, when ballots containing all the qualified candidates—basically any movie that came out during the year and played in Los Angeles for at least fifteen days—are sent out to each member of the Academy. Members pick the five people they think have done the best work in their branch that year, and also mark their choices for Best Picture. Polls close on February 1. After the nomination forms are collected, they are delivered to Price Waterhouse, an accounting firm, to be tabulated. The top five vote-getters in each category are placed on the final ballots, which are sent out to members at the beginning of March. During the next two weeks, every movie on the final ballot is screened twice so that voters will have a chance to see the work before the polling deadline. As with the nomination, each category in the final election is broken down by Academy branch: members of the actors branch vote only in the acting categories, and so forth. Price Waterhouse counts the votes and the winners are announced, with much hoopla and speechifying, at the made-for-TV ceremony near the end of March.

The peer-based nomination process sounds perfectly credible, but there are a couple of big problems. First, voter turnout is worse than in general political elections, averaging no better than 60 percent in good years, and often dipping as low as 40 percent. By default, control of the voting process is often left to studio flaks, who do everything to carve their employers a share of the Oscar pie. Second, the Academy is a backward-looking, Old Guard sort of institution, many of whose members are resistant, or oblivious, to innovations in the movie-making business. (A person becomes an Academy member after being nominated by two current board members and approved by a vote of the board of directors.) Of the directors, 70 percent are older than sixty, 45 percent have not made a movie in ten years, and 20 percent have not made a movie in twenty years. Cutting edge they're not. Also, this being Hollywood, backbiting is endemic. One of the most celebrated cases in recent years was the treatment received by Steven Spielberg in 1985, when his picture *The Color Purple* garnered eleven nominations but he was left off the Best Director ballot. If the movie was that good, indignant critics railed at the time, *someone* must have been responsible, and it was probably the director. Indeed, Spielberg had already won that year's Directors Guild Association award. According to widespread rumor, the Academy was turned off by Spielberg's cocky attitude and not a little jealous of his ever-growing fame and fortune.

The Academy's credibility problems are almost as old as the institution itself. Founded in 1927 to generate publicity for the huge studios, the Academy quickly grew from fewer than a dozen members to 1,200 in 1932. During the early thirties, omnipotent producer Louis B. Mayer, in an attempt to freeze legitimate labor unions out of Hollywood, tried to turn the organization into a studio-controlled union. Members began resigning in droves, and by the time Frank Capra became president, in 1935, the ranks had dwindled to forty. The Academy has expanded steadily since then, but it has never outgrown its penchant for bestowing its awards on safe movies. Among the dubious Best Picture winners are *How Green Was My Valley,* which beat *Citizen Kane* in 1941; *Patton,* which topped *M*A*S*H* in 1970; and *Rocky,* winner over *Network* in 1977.

?

How do they make nonalcoholic beer?

LEGALLY SPEAKING, in the United States there is no such thing as nonalcoholic beer. According to government definition a nonalcoholic drink is a beverage that contains no alcohol of any sort. The law also maintains that a beverage containing less than 0.5 percent ethanol cannot be called beer. The end result is a semantic issue, which brewers of alcohol-free drinks that look like beer, come in six-packs like beer, smell like beer, and taste more or less like beer, resolve by calling their products "nonalcoholic malt beverages."

Nonalcoholic beverages and low-alcohol beers have been around since Prohibition, but no one ever mistook them for the real thing and they remained largely shunned. In recent years, however, a surging national awareness of health and fitness has created a viable market for alcohol substitutes, and now brewers are applying new technologies to the quest for a really good, beery, nonintoxicating brew.

There are two ways to make nonalcoholic malt beverages. They can be specially brewed to avoid alcohol, or they can be brewed as normal beer, with the alcohol removed later. However it is done, something must be used to compensate for the missing alcohol, which adds to both the flavor of beer and its characteristic foaminess. Beer is made by steeping a mixture of barley and hops, called mash, in hot water to produce wort. The liquid wort, high in organic sugars and enzymes, is caused to ferment by adding yeast to it. The fermentation process turns sugar into alcohol and carbon dioxide, producing the foamy, slightly bitter drink enjoyed the world around.

There are a few basic ways a brewer can prevent alcohol from forming in the wort, and thereby brew a nonalcoholic beverage. He can use a weak, low-sugar wort, devoid of the ingredients yeast feeds on to produce alcohol. He can refrain from adding yeast. Or he can add yeast but do so under conditions, such as low temperature and excess CO_2, that make fermentation impossible. The last method works particularly well because it uses normal yeast and wort, and the yeast enhances flavor by reducing noxious by-products in the wort while leaving desirable ingredients—glucose, fructose, maltose and amino acids—intact. Flavor extracts and stabilizing agents are also used sometimes to duplicate the taste and feel of real beer.

Removing alcohol from fully brewed and fermented beers is becoming a popular alternative to preventing fermentation. The product, linguistic wrangling aside, is beer minus the alcohol, spiritually akin to decaffeinated coffee. The most effective way to remove alcohol from beer is by vacuum distillation. The fermented beer is warmed and then pumped into a vacuum chamber and condensed, a process that separates the alcohol from the body of the beer and reduces the beer to a thick concentrate. The concentrate is then cooled and water and CO_2 are added to produce a carbonated malt beverage. Vacuum distillation is advantageous because in removing the alcohol it avoids the extreme high temperatures that damage color and flavor.

Another way to remove alcohol from beer is by something called the "ultrashort time evaporation," or Centri-therm process. Beer is placed in a centrifuge and made to pass over a heated surface, which causes the alcohol to evaporate. The process damages yeast and removes some flavor compounds, but since the beer is in contact with the heating device for only about half a second, no significant harm is done.

Alcohol can also be filtered out of beer through a semipermeable membrane. The beer is placed under great pressure so that it breaks down into its component molecules. Ethanol and other small molecules are able to escape through a membrane that retains the larger, desirable molecules. In essence, filtration squeezes alcohol out of the beer, leaving behind a concentrated solution that is diluted with water before being bottled and sold as a "non-alcoholic malt beverage."

Most of the upscale pseudo-beers, the ones that masquerade as imported ales in the liquor store cooler, and taste pretty close, are made by a vacuum distillation alcohol removal technique.

?

How do they know there was an ice age?

THE MOST OBVIOUS CLUE to ice ages past is that in some parts of the world, things seem markedly out of place. This is how, in the late eighteenth century, Swiss peasants discovered that there had been ice ages past. Noting that glaciers in the Alps were slowly transporting and depositing rock masses down-valley, they correctly inferred that boulders strewn in their pastures had arrived in similar fashion centuries earlier. They also noticed that the polished and scratched or finely grooved and rounded bedrock knobs along the valley walls and floors resembled those surfacing from beneath the melting ice of existing glaciers. This similitude hinted that the knobs had been produced by moving glacial ice during a time when the glaciers extended farther down the mountain valleys.

Similar discoveries were made in Germany and Scandinavia. Eventually geologists concluded that huge sheets of moving ice had traveled from the far north to the plains of northern Germany. They deduced, furthermore, that the world had experienced a period of glaciations that had occurred and would recur in 10,000-year cycles, each glaciation lasting some 100,000 years. They called this vast period—in which we are still living—the Great Ice Age.

During the Pleistocene Epoch, which began about one and a half million years ago, mountain glaciers formed on every continent, and vast glaciers, as thick as several thousand feet, sprawled across northern North America and Eurasia. These far-reaching glaciers intermittently covered nearly a third of the earth's present land surface. Today, remnants of the great glaciers encase almost a tenth of the planet, demonstrating that conditions somewhat similar to those that spawned the Great Ice Age are still operating in polar and subpolar climates.

The first Ice Age studies in North America began in 1846, when Swiss geologist Louis Agassiz arrived in the United States. Early analysis of this country indicated that large parts of the north central and northeastern United States had once been blanketed by thick ice sheets. Traces of glacial erosion and depositation were ubiquitous: rock outcrops were smoothed and polished, scratched or striated; hills were rounded and wrapped with glacial debris; valleys were smothered by sand and gravel deposited by glacial melt waters. All these conditions were the result of the slowly advancing ice sheets, which had sucked up soil and loose rock, plucked and gouged boulders from outcrops, and hauled this clutter elsewhere, often great distances away. En route this hard material served as an abrasive—smoothing, polishing, and scratching all rock outcrops in its path, thus producing a more subdued landscape with rounded hills and higher valleys.

When glaciers melted, blends of clay, sand, gravel, and boulders were deposited as an unconsolidated mantle on the countryside. Thus, the once rolling hills of the Midwest are now level, and great river valleys such as the Mississippi and Ohio, overwhelmed by ice sheets, were forced into new channels. Ancient Lake Bonneville, the largest glacial lake in the western United

States, once covered more than 20,000 square miles and was more than 1,000 feet deep. The Great Salt Lake in Utah is but a puddle compared to its original size.

Although the first appearance of man remains unknown, scientists attribute our rapid physical and cultural development during the Great Ice Age to the necessity of adapting to the changing climates of that time. Our most primitive tools and skeletal remains from ancient deposits are concurrent with periods of glaciation in Africa, Asia, and Europe. The melting of the immense ice sheets sparked the evolution of the Bronze and Iron Age cultures. And animals suited to cooler climates became extinct, particularly larger mammals such as the woolly mammoth, mastodon, dire wolf and saber-toothed tiger.

?

How do they correct nearsightedness in ten minutes?

IN 1974, a Soviet ophthalmologist named Svyatoslav Fyodorov developed a surgical procedure that has proved a speedy and sometimes very effective method for correcting nearsightedness. The basic idea is to alter the curve of the cornea, which is the window into the eye, by means of tiny surgical scars. Performed under local anesthesia, the procedure, called "radial keratotomy," usually takes a half an hour, but some doctors claim to do it even faster. It costs about $2,000 for each eye, and the patient can go back, with a patch and some discomfort, to what he was doing that day. Here's how it works.

A person with nearsightedness sees blurry images because his cornea is too convex, causing light rays to focus in front of—rather

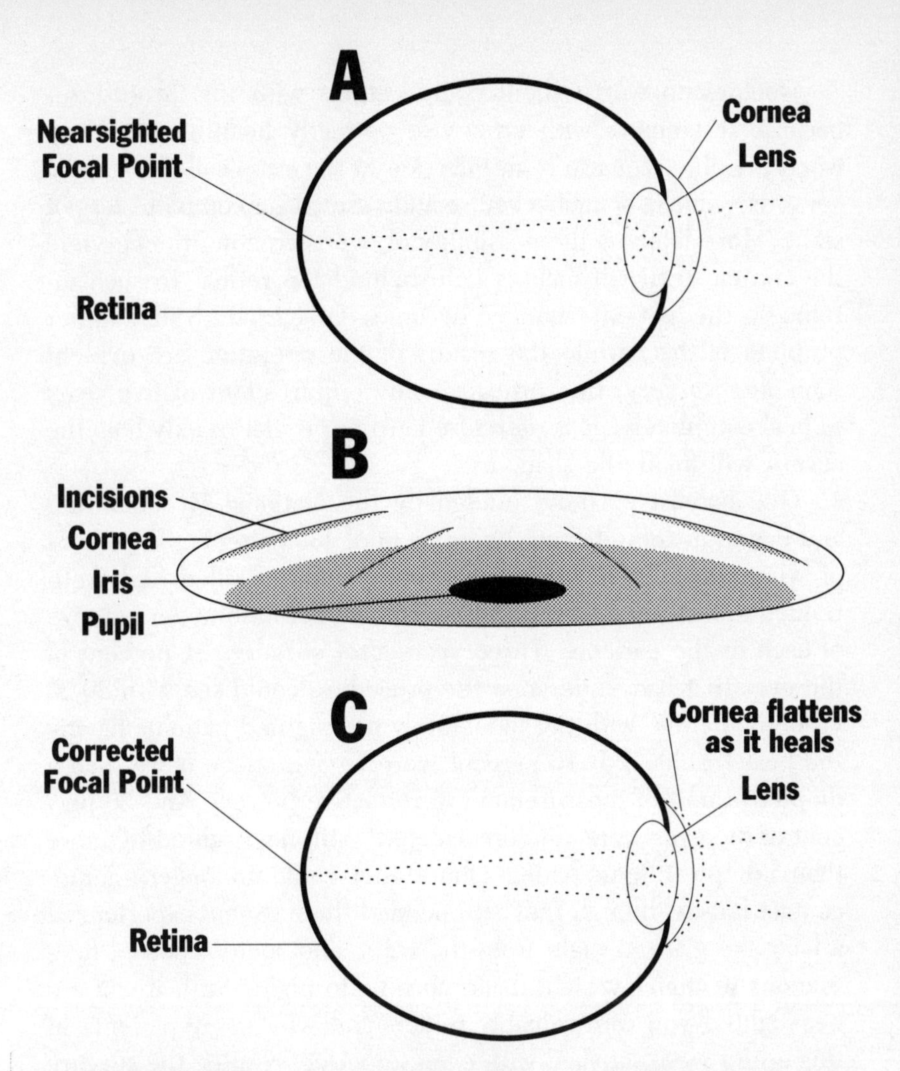

In a nearsighted person, the uncorrected eye (A) focuses images in front of the retina. To correct the problem, an opthalmologist cuts a tiny starburst on the cornea of the eye (B). As the wounds heal, the cornea flattens. The flatter cornea (C) focuses images on the retina, where they should be for clear vision.

than *on*—the retina in the back of the eye. To correct that condition, an ophthalmologist uses a diamond-bladed knife and a microscope to make four to sixteen very small incisions that look like a perfect star burst on the cornea. As the tissue heals, the cornea flattens, so that without the aid of glasses or contact lenses it now brings light rays to a focus where they should be.

Doctors do worry about complications with this procedure, because it tampers with otherwise perfectly healthy eyes. The worst possible scenario is an infection in the scars called bacterial keratitis, which, if unchecked, could result in a complete loss of sight. More likely is the possibility of overcorrecting the curve of the cornea so that it focuses light behind the retina. In such instances, the patient actually becomes farsighted. Still another problem is that, while the results of the operation are evident soon after surgery, the cornea actually requires four or five years to heal completely. It is therefore hard to predict exactly how the cornea will finally be shaped.

One extensive study, funded by the National Eye Institute and begun in 1980, tested the eyesight of 435 patients, 99 percent of whom had 20/50 vision or worse. Highly skilled and well-trained ophthalmologists performed radial keratotomy on one eye of each of the patients. Three years after surgery, 51 percent of the eyes that had undergone the procedure could see with 20/20 vision or better, with the moderately nearsighted patients having the best results. But 16 percent were farsighted by more than a diopter (a unit of measurement of refractive power). And 26 percent of the eyes were "undercorrected": still nearsighted by more than a diopter. Some patients found they could no longer tolerate contact lenses, though they still needed them. Some experienced a halo-like glare at night from the scars, and some reported fluctuations in their eyesight from morning to night. Still, if you are nearsighted you can probably understand why many patients in this group were satisfied with even imperfect results; the surgery left them less dependent on corrective lenses than they had been before. About three quarters of the whole group opted for the same surgery on the second eye.

A new technique using a computer-controlled laser beam instead of a diamond knife is being explored. "Photorefractive keratectomy" may be a more precise method than its forerunner, but the jury is still out. In either case, some doctors are suggesting that people wait a few more years before ditching their glasses, until both techniques are better known.

?

How do lawyers research members of a jury before deciding whether to challenge them?

JURY SELECTION ISN'T just based on instinctive hunches. As the stakes of cases have increased, lawyers on both sides of a case are carefully weighing each potential juror and trying to eliminate those who may have predispositions against their client. It is a tricky task requiring rapid psychological analysis by lawyers, assistance from jury consultants, computer character profiles, and carefully crafted questions.

In all courts, judges have tremendous discretion about how the selection will proceed, and lawyers in the court need to familiarize themselves ahead of time with the particular judge's modus operandi. The selection in federal courts is generally quick, often under an hour. Each side is permitted only three peremptory challenges; that is, the counsel for the defense and the prosecution can each strike only three prospective jurors without providing a reason. (However, recent Supreme Court cases have limited this right somewhat.) Both sides are also allowed unlimited challenges for cause, but here the lawyer must provide a reason for eliminating someone, usually tangible evidence that that person cannot be unbiased. Federal judges tend not to dismiss candidates for cause, however, and that helps expedite the process. Every state court has different policies about the length and type of questioning permitted. The process can take days, especially if the case is to be a long one in which juror alternates will be needed.

What do lawyers look for and how? "The trial lawyer can't just glance or casually look at a member of the jury panel," explains William H. Sanders of Blackwell Sanders Matheny Weary and Lombardi. "The trial lawyer has to *eyeball* each and every member of the jury panel. Just as a portrait painter sees so much in his subject's face, the trial lawyer must scrutinize the body language, demeanor, expressions and dress of each individual." During the seconds before the candidate is seated, the lawyer watches his or her stride—is it forceful or lazy? How is the candidate dressed—high heels or running shoes? Is the candidate carrying any reading material, and if so, what kind? The lawyer is looking for clues about how this potential juror will view his client. Does he or she appear liberal or conservative, rigid or easygoing, sympathetic or not to the plight of his client?

After eyeballing the prospective jurors, the lawyer may pose a series of questions (or in some cases have the candidates fill out a questionnaire) about each candidate's occupation, marital status, age, and residence. Usually the lawyer is not allowed to inquire about religious affiliation or political leanings. Inevitably the lawyer will rely somewhat on occupational stereotypes. For example, in a case of wrongful death of a child, a young mother who will sympathize with bereft parents will be more desirable for the defense than a seventy-year-old widow. But the lawyer must also try to read between the lines. Is this young mother an angry, frustrated person? How does she feel about her work, her kids? He must reach rapid conclusions drawing on general demeanor and personality type.

If a trial lawyer is challenging for cause, he does so at the close of the questioning, or voir dire. To obtain a reason he can ask leading questions. Say the case involves a drunken-driving accident and a potential juror professes to be a teetotaler. The lawyer would highlight the candidate's preference for abstinence while inquiring whether he could indeed be fair to the defendant; wouldn't he be likely to award more damages against the driver because of his drinking? If the lawyer can demonstrate to the judge that the candidate is prejudiced, then that person will be disqualified.

In big cases lawyers often hire jury consultants who study social patterns and personality types for insight into what motivates an individual. Some jury experts try to read personality traits through somatotypes, or body types. The endomorph, for instance, who is obese, tends to be free and direct in answering questions. His generosity makes him a desirable juror for plaintiffs in civil cases. At the opposite end of the spectrum is the terribly thin ectomorph, who is considered anxious, rigid, tight with money, a juror for the civil defendant.

Some lawyers use personality typing by psychiatrists in the selection of a jury. Take the obsessive-compulsive—a motivated, meticulous achiever and go-getter. "That's the person I don't want for the plaintiff," writes lawyer Bill Colson in *Jury Selection: Strategy and Science*. "That person really thinks that you work your way to heaven—you should never get anything for nothing and that we can always overcome. So for me to stand up and say that I want him to give some money for intangibles, for pain and suffering or for future disability or for a scar on a body or a burn—that just wouldn't enter his mind."

Other experts classify prospective jurors in five social categories: leaders, followers, fillers, negotiators, and holdouts. Lawyers generally don't like leaders, whose independence makes them hard to sway. If, however, the lawyer is sure a leader is on the "right" side, then he or she would be a good pick. A lawyer who has found such a person would then try to put as many followers on the jury as possible. That's because followers, the passive ones more accustomed to receiving orders than giving them, are likely to fall in line with the leaders, or strongest members of the jury. What lawyers call filler types might also be desirable. Because fillers display indecision but not outright resistance, any lawyer worth his salt will think he can sway them to his client's cause. Of course, a prosecutor or plaintiff's lawyer would never want a holdout, a person unlikely to go along when the majority of the jurors want to convict or find for the plaintiff. The lawyer must carefully weigh, then, not only each individual but the dynamics of the group. And stereotyping, however useful, isn't foolproof.

To help make jury selection more scientific, Decision Research, a market research firm with a strong litigation service practice in Lexington, Massachusetts, stages mock trials in which a "jury panel" is presented with the facts of a "hypothetical" case. Observing through a one-way mirror, lawyers and Decision Research analysts study how people react to the arguments and lawyers. Says Robert Duboff of Decision Research, "You start to see fairly predictable patterns." In a euthanasia case, for example, he found that people under forty and over sixty were more likely to sympathize with the defense. Those in between, in the forty- to sixty-year-old range, were the least sympathetic. "Possibly they felt some trepidation that someone would do that to them when they're not ready to go," suggests Duboff. "But cases are usually not that clear-cut." When the data are more varied and complex, Decision Research does computer runs to establish patterns. In general, it seems inadvisable in a criminal case for the defense to select a jury too much like the defendant. (Recall the Scarsdale Diet Doctor case in which Jean Harris was convicted for murder by a strategically selected bunch of middle-aged, middle-class women.) "If you are similar to the defendant, you judge him or her by your own standards," says Duboff, "and you say to yourself 'I'd be tempted to do that, but I wouldn't,' so you want to punish him."

Despite all the computer analyses, surveys, and studies, every time a lawyer steps up to a panel of jurors he is, in the end, dealing with unpredictable individuals, and their responses may hinge on unperceived quirks and foibles. Even so, it can't hurt for lawyers to try to ensure the scales of justice aren't tipped against their client.

?

How does an electric eel generate electricity?

IN THE WARM, murky waters of the Amazon and Orinoco Rivers of South America lives a sluggish fish with a shocking capability. The electric eel can emit an electrical discharge of 550 volts. In the animal kingdom, fish alone can produce electricity, and of the 250 species that do so *Electrophorus electricus* is the most powerful. It can deliver a nasty shock to a man ten feet away or stun other fish, which it then consumes. According to ichthyologist C. W. Coates, "Any average electric eel of three feet or longer has a power pickup of about 150 horsepower per second, a fact which seems to belong in the 'Believe It or Not' class."

The electric eel usually reaches four to eight feet long. It moves stiffly by undulating its long anal fin, which runs from the throat to the tip of the tail. All essential organs are located in the front fifth of the body. The rest houses the tail musculature and the electrical works, consisting of two small "batteries" and a larger one. These organs, which derive from muscular tissue, are made up of specially coated units called electroplates. Seventy columns of electroplates run along the sides of the electric eel's body, and each column contains from six thousand to ten thousand plates. The electroplate surfaces face the head and tail, all positive poles lying one way, negative poles the other. Electricity flows from the tail to the head.

Cranial motor nerves power the electric organs. Impulses race through the eel's body twelve to fifteen times faster than the speed at which impulses travel from the human brain to the nerves. The electroplates each contribute 150 millivolts; connected in series, they combine to produce a remarkably high voltage.

The first small battery in the electric eel's tail is operating all the time. As the eel moves, this battery emits low-voltage waves that set up an electric field in the water. When an object enters this field, it causes a distortion, which is registered by pits in the eel's head. In this way the eel apparently perceives the size and nature of things around it; even with poor eyesight, the electric eel successfully finds food in the murkiest of water.

Scientists believe that the second small battery ignites the larger battery, which in turn discharges a series of quick, powerful pulses. Each pulse lasts three milliseconds and is delivered with incredible speed. So fast is the pulse that, even though its voltage is more than sufficient to light up many light bulbs, it would come and go before the inertia of an ordinary light bulb was overcome. The lamp would require one fifteenth of a second to go on, whereas the rapid-fire discharge of the electric eel pulses and passes in one tenth that time.

?

How does an invisible fence keep your dog from straying?

HOW DO YOU KEEP Fido faithful to the homefront in the face of overwhelming temptation—a female trotting by or the cat next door? Traditional fencing is pricey, and furthermore, some dogs dig under a fence or even scale the chain-link type. An alternative solution is invisible fencing.

The system consists of a thin antenna wire of whatever length you need to circumscribe your property, or the area in which you wish to contain your dog. The wire is buried one to three inches underground, and both ends are attached to a radio transmitter placed, say, in your garage or in your home. The transmitter emits a radio signal that is picked up by a receiver in a small rectangular box affixed to a special collar worn by the dog. The receiver, which is positioned under the dog's neck, has two posts (available in short or long) that extend from the box to the dog's skin. When Fido approaches the wire outlining his terrain, he hears a warning beep. The depth of the signal field, or area on either side of the wire that will activate the collar, is usually six to eight feet but can be varied to fit your particular property. If the dog ignores the sound, keeps going, and crosses the wire, it receives a mild shock, the strength of which can be adjusted to meet the level of your pet's doggedness. The receiver is powered by a 7.5-volt battery. No hazardous electricity is involved, so the system is safe for households with small children.

One wonders how the poor dog can guess where it's safe to wander without getting zapped. The training process is crucial. According to the Invisible Fence Company, Inc., based in Wayne, Pennsylvania, which began marketing the system in 1979, many dogs can learn their boundaries after receiving only a single, harmless shock. The system relies on a technique first discovered by the Russian physiologist I. P. Pavlov in the early 1920s—the dog's susceptibility to being conditioned. The dog can learn a particular response to a particular stimulus. In the case of Pavlov's dogs, a ringing bell first associated with food evoked a certain response, and continued to do so even later when no food was presented. In the case of the invisible fence, the dog becomes conditioned to step back the instant it hears the beeper on its collar.

When you first install an invisible fence, its parameters are marked by clearly visible flags placed six feet apart and about three feet inside the loop wire. The dog should be at least five months old and used to walking on a leash. Lead the dog to a flag, shake the flag and say "NO!" when the warning beep sounds.

Then let the dog walk on its own into the signal field and feel the correction. One hopes that once will suffice and that the dog will take note of the flags and the beep thereafter. When the dog returns to the safe area, praise it. You must work with the dog about fifteen minutes each day for a week. During the second week the flags are gradually removed and Fido can be trusted to play outdoors unattended.

Although invisible fences are popular and largely effective, some dogs persist in breaking through them if the call of the wild is strong enough. Another drawback is that even if your own dog is obedient, an invisible fence does nothing to prevent neighborhood pets from trespassing and disturbing dear Fido—and you.

?

How does a thief break into your car and drive off with it in less than a minute?

ALL TOO EASILY, according to the New York FBI Auto Theft Task Force. It should know. In the city with by far the largest number of vehicle thefts in the nation, the FBI grapples with the large-scale professional syndicates that enjoy most of the profits of this lucrative crime. Meanwhile, its task force vehicles are repeatedly stolen when parked in front of its headquarters.

The techniques of the professional thief are so slick that it takes him an average of only sixty seconds to break into a car, no matter how well locked and alarmed. As a result, a motor vehicle is stolen somewhere in the U.S.A. every twenty-two seconds, adding up to about a million stolen nationwide each year. About 133,000 were swiped in 1989 in New York City alone. And the rate continues to increase.

Just how does a thief get into your locked car? He may poke the lock or pry down the windowpane with a "slim jim," a long, very thin metal tool that can be slipped down between a car's windowpane and door frame to reach the door's lock mechanism. In a matter of seconds the thief is able to move a bar on the lock and open the door.

Once in, the thief can start the car without its key by using one of a number of devices readily available to him on the open market, through crooked suppliers, or through theft. For instance, he may have a supply of master keys, and if those are insufficient, a portable key cutter. Certain makes of car have their key code numbers engraved inside the glove compartment or passenger door lock. A thief can pull out a door lock with pliers, find the key code number, check in the manufacturer's manual to see what corresponding key to make, and cut it on the spot with the portable key cutter in a matter of moments. Failing that, there are a number of tools, such as dent pullers (also known as slide or slap hammers), that can start a car instead of a key.

Since the value of a luxury car such as a Mercedes-Benz decreases considerably if damaged—as it is quite likely to be if broken into by such techniques—car thieves often steal them *with* their keys by entering parking garages and bribing an attendant. They also may take a car at gunpoint when its driver is stopped at a red light, or bump into it with another car from behind and then flee with the expensive car when the owner gets out to examine his fender.

Car alarms are no deterrent, since thieves can readily get hold of directions explaining how all available alarm systems can be cut off. Experienced thieves can fool even the most elaborate alarm systems of expensive cars with a meter that can be bought for $150 from any electrical store. It analyzes the diodes in the electronic brain of the alarm system so the thief can then electronically simulate the right key with a resistance diode.

There is little you can do to protect your car against a thief who wants to make off with it. Concealed cutoff switches that have to be turned on to start the engine sometimes prevent thieves from making a rapid getaway but generally not from taking the car. Steering column locks can be broken, and marking your car's VIN (vehicle identification number) on all its major strippable parts does not deter chop shop operators, since they can conceal even deeply etched numbers.

According to police estimates, the "joy rider," or casual thief, accounts for only a small percentage of vehicle theft, although what this precise figure is is not known. A surprisingly larger proportion (25 to 30 percent) is made up of ostensibly solid citizens who want to get a new car out of their insurance company by arranging the "theft" of their present one, often in collaboration with corrupt auto repair or chop shops. Most vehicle thefts, though, are carried out with professional efficiency and speed by organized criminal groups whose hefty profits add up to big business and contribute largely to the over $4 billion annual loss sustained by the nation. Car theft is second only to drugs as organized crime's biggest money earner, and the two activities are closely linked.

?

How do they get the oat bran out of oats?

YOU DON'T GET THE OAT BRAN out of oats, but *off* of oats. The bran is the husk, or outer covering, which is removed during the milling of oats. First the oats are cleaned to remove foreign matter and inferior specimens. Then the grain goes through a drying and roasting process to reduce their moisture content. This process also develops flavor, helps prevent spoilage, and facilitates the subsequent separation of the hull from the groat, or internal grain. After the oats are cooled by circulation, they are ready for the hulling process.

The hulling machine separates the groat kernel from the surrounding hull and produces the following: hulls, groats, broken groats and meal, flour and unhulled oats. These materials are separated by air aspiration and sifting: they are placed in rotating, meshed, cylindrical frames and the hull (bran) is either sifted or flung out of the frame. It's a little like panning for gold. In this case, though, what used to be discarded or used primarily for animal feed is now regarded as valuable for humans.

In the early sixties scientists discovered that oat bran could significantly lower cholesterol levels in human beings. Other grains, such as wheat, contain only insoluble fiber, which passes right through the body. Only oat bran has both insoluble and soluble fiber. Soluble fiber gets absorbed into the bloodstream and helps remove cholesterol from inside the arteries and prevent it from being deposited on the artery walls. Moreover, the cholesterol that is decreased is the harmful LDL cholesterol. Oat bran does not affect levels of HDL cholesterol, which actually helps *prevent* clogging of the arteries.

With the increasing public awareness of these findings, oat bran has become the "miracle" ingredient in cereals, muffins, bread and even doughnuts. According to studies conducted at Northwestern University Medical School (funded by Quaker Oats Company), the average person needs to eat some thirty-five grams of oat bran or oatmeal every day to lower his cholesterol by about 3 percent. One ounce of pure hot oat bran cereal comes close to fulfilling that requirement. But you need to eat your oat bran pure and unadulterated to derive the maximum benefit. As it is, the benefit of the bran is lost in, say, an oat bran muffin whose unwholesome fat content may outweigh the healthy qualities of the bran. If you are buying oat bran specifically for its cholesterol-lowering capacity, you should always check the label to see what, besides the bran, you are buying.

?

How do they get the bubbles into seltzer?

AN ENGLISH CHEMSIT by the name of Joseph Priestley bears the distinction of having invented seltzer water, in itself a popular beverage and the precursor of the ubiquitous soft drink. One day in 1772, Priestley mixed up a solution of baking soda, vinegar, and water, a combination that produced carbon dioxide, which created a fizzing action as the gas escaped from the water. Priestley's experiment was regarded highly enough to win him the Copley Medal of the Royal Society in 1773, but a hundred years would pass before seltzer water caught on as a refreshment, when a beverage called Lemon's Superior Sparkling Ginger Ale became the first registered soft drink.

Carbon dioxide is still the agent of soda water's bubbles, but instead of relying on baking soda to get CO_2, modern seltzer bottlers inject CO_2 gas itself directly into the water.

Carbonation is a two-step process. During the first step, CO_2 is injected under pressure into water. From two different storage containers, pure water and CO_2 are pumped into a common refrigerated, pressurized tank, called the sparging unit. The water absorbs the CO_2, but only up to a certain point; after the water has reached its saturation point, the bubbly solution is pumped into a second pressurized tank, the carbonating unit, for the second stage of carbonation. A trough at the top of the second tank disperses the liquid into droplets, which will still absorb gas, and more CO_2 is introduced until the water is fully carbonated. The seltzer is pumped from the carbonating tank into a filling machine, which disgorges it into individual bottles and cans.

***FACING PAGE:* Seltzer is made by combining water and CO_2 in a tank. After the water is saturated with gas, it is pumped into a second tank and broken up into droplets, which absorb additional CO_2.**

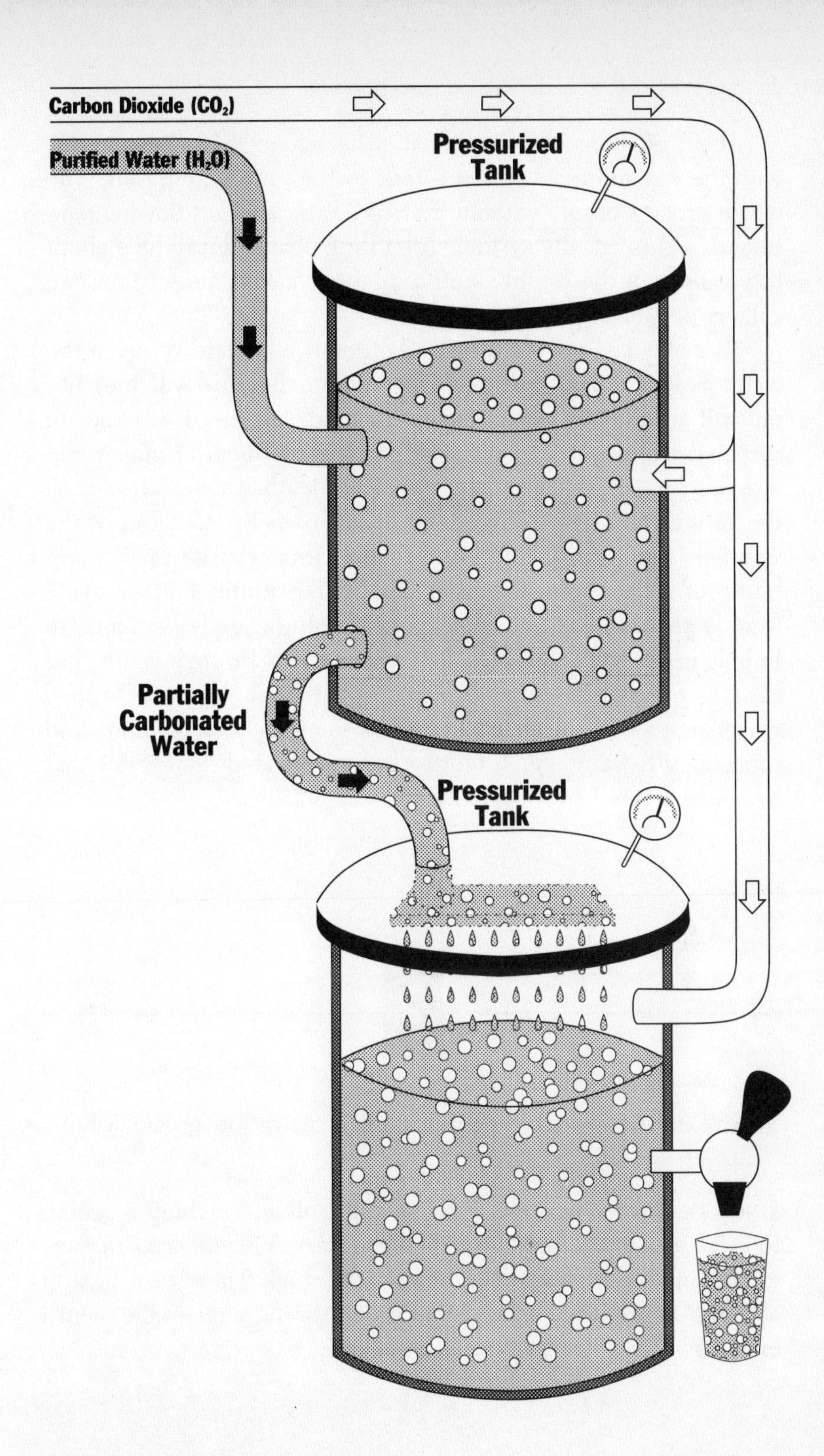

Carbon Dioxide (CO_2)
Purified Water (H_2O)
Pressurized Tank
Partially Carbonated Water
Pressurized Tank

About 80 percent of carbonation is achieved in the sparger, with the remaining 20 percent done in the carbonating unit. The whole process occurs without a break, water and gas flowing continuously through the system. A bottling plant can produce about fifty gallons of the bubbly stuff a minute, making three thousand gallons per hour.

To make flavored soft drinks, bottlers simply add syrup to the water as it is pumped through the system. Because a lighter liquid will absorb CO_2 more readily than a dense one, it is easier to carbonate pure water than a soda pop like root beer or ginger ale. It is also easier to carbonate a cool liquid than a warm one, so the various tanks are refrigerated: heat drives the CO_2 out of the solution. You've probably noticed that when you uncap a warm bottle of soda water, it gushes like Old Faithful. Seltzer made from water that has not been meticulously filtered also tends to bubble excessively. Bubbles do not float freely through water, but cling to naturally occurring microscopic particles. "Hard" water, which has a high alkali count, offers more bubble toeholds and consequently fizzes much more when carbonated than water that has been treated to remove alkali.

?

How do they teach guide dogs to cross at the green light?

SINCE DOGS ARE BELIEVED TO BE color blind, teaching a canine to cross at a green light sounds like a very difficult task. In fact, guide dogs don't look at the traffic light at all. They learn to stop at curbs so that their master may listen carefully for traffic before crossing.

Training a guide dog is a lengthy and serious process, one that can be undertaken only by professionals. It is done at any one of ten private, nonprofit schools, such as the Guide Dog Foundation in Smithtown, New York, that teach blind people and dogs to work together as teams. This type of dog is sometimes called a Seeing Eye dog. The term comes from another private school, Seeing Eye, Inc., in Morristown, New Jersey, which holds a trademark on the name.

Before the schooling starts, the dog is meant to enjoy a happy, well-rounded puppyhood. The Guide Dog Foundation breeds its own stock of Labradors and golden retrievers, which are farmed out to foster families, known as "puppy walkers." The family's primary responsibility is to care for the dog and expose it to as many different situations as possible. It is expected that the dog will be taught to walk nicely on a leash, sit, stay, lie down and come when called, but nothing more. When the puppy reaches a year old, it returns to the Foundation and begins its training, which will last from three to six months.

One of the first things the dogs learn is to walk straight down the center of the sidewalk without sniffing or being distracted by people or other dogs. Elane Siddall, an instructor at the Guide Dog Foundation, explains that this is done with positive and negative reinforcement: "You're constantly telling the dog, '*Straight*, come on now, *straight*,' and when the dog's doing well you praise it—'Good dog.' When the dog does something inappropriate, you give a snap on the leash, '*No*, now *straight*,' and give it something immediately to counterbalance its correction, something you can praise it for."

Once the dog has learned to walk straight down the center of the sidewalk, deviating only for obstacles, it is taught to stop at curbs. Siddall does this by approaching the curb while repeating to the dog, "*Straight* to the curb." When she is about six feet from the curb, she says, "*Steady* to the curb," slowing down the dog as she gives the command. If all goes well, the dog realizes that something is changing. About two steps from the curb, Siddall says, "*Halt*," and gives a slight jerk on the leash and a release. When she has stopped the dog, she pats the curb with her foot and says, "*Good* curb, *stay* at the curb, *good* curb." After many repetitions, the dog eventually begins to stop at the curb on its own.

Then it will be up to the owner to listen and determine when to cross. When he hears traffic start up at his side, he knows he can safely cross the street. If the traffic in front of him is moving, he knows he must wait.

If a blind person makes a mistake and winds up in an intersection with a car coming, the dog will either halt until the car passes, pull its master to the side and out of the way, or put itself between its master and the obstacle. When the car has passed, the dog will lead the blind person forward to the opposite curb.

?

How do they measure dream activity?

IN THE EARLY DECADES OF the twentieth century many scientists and philosophers speculated upon the nature of our dreams but could not study them scientifically. Researchers could wake up volunteers at different times in the night and ask them whether they had been dreaming and for how long. But the propensity of dreams to vanish into oblivion almost as soon as we wake made for such sparse and unreliable results that dream activity, however intriguing, was held unsuitable for scientific study.

In 1937 scientists began studying sleep with the help of an electroencephalograph (EEG). This equipment, invented in 1929, can record very small fluctuations in electrical activity, commonly known as "brain waves," constantly occurring in our brains. Five types of electrical activity were identified as taking place in recurrent sequences during a night's sleep. Then in 1953, a University of Chicago graduate student, Eugene Aserinky, discovered distinct periods of rapid eye movement (REM) during babies' sleep. Scientific research into dreams took off in earnest.

Researchers found it quite easy to differentiate the periods of rapid eye movement (which can be detected under the closed eyelids of sleeping adults as well as those of children) from the eyes' immobility or slow rolling movements (non–rapid eye movement, or NREM) during the rest of sleep. Brain wave rhythm during REM sleep has been found to be more like that during the waking state than the other sleep states, and to recur regularly about every ninety minutes throughout the course of a night's sleep. So sleep was reclassified as NREM during stages 1 through 4 and REM in stage 5.

The possibility that the rapid eye movements might correspond with sleepers' watching their dreams prompted researchers to wake up their subjects during REM and NREM sleep and compare the results. They found that about 80 percent of the time that sleepers were awakened from REM sleep, they reported having been dreaming, whereas only 7 percent of the time did sleepers aroused from NREM sleep say they were in the midst of a dream. These findings led to the still continuing era of the sleep laboratory.

A research subject, often a psychology student earning experimental credits, goes to bed for a night's sleep in the lab with about eight electrodes applied to certain places on his head. The EEG amplifies the tiny shifts of electrical potential occurring within the brain and body that are detected by the electrodes. Continuous shifts from positive to negative charge within the subject's brain cells are transmitted to magnetic tape, which drives a row of pens, each of which reflects changes in a different part of the brain or body. When an electrical charge is negative, a pen will swing upwards to a corresponding magnitude; when positive, the pen will swing down. The height of the waves shows how much voltage is being generated by the area studied, and the distance between the waves indicates the rhythm and rapidity of electrical movement in progress.

In order to compensate for the distorting effects of sleeping in a laboratory, researchers commonly document their subjects' sleep patterns over five to ten consecutive nights in order to establish a normal pattern. Still, the possibility of a certain amount of "laboratory effect" inhibition is tolerated as unavoidable. Nightmares are rare, for instance, as are nocturnal sexual emissions in the lab. Some dreamers vigorously deny any violent or sexual content in their dreams, despite strong indications of agitation of this nature to clinical observers.

The experimenters monitor the sleeping subject as he progresses through stages 1 through 4 of (NREM) sleep, delineated by their different brain wave patterns. When, after about ninety minutes, the subject shifts into stage 5 (REM) sleep—presaged by a highly irregular, rapidly changing but low-voltage brain wave printout very similar to that of the waking brain—the dream researchers' attention focuses on the eye movement printouts (measured by two particular electrodes) on the polygraph paper. The script is now characterized by abrupt shifts, as the sleepers' eyes dart to and fro as though looking at a movie. If the subject is woken up at this point, the odds are ninety-nine in a hundred he will say he was dreaming. But the experimenters will probably let the subject dream for about five minutes, the approximate length of the first dream phase of the night, before they awaken him to discover details of the dream. Meanwhile, monitors recording data about the subject's physical state show an increase in oxygen consumption and internal brain temperature and irregularities—wild fluctuations even—in heartbeat and respiration, similar to those during violent emotion or exertion when awake. Although the dreamer may grind his teeth, the rest of his muscles are completely slack.

After being awakened briefly, the subject will be allowed to fall back to sleep. Again he will drift down through stages 1 through 4 of NREM sleep and, in ninety minutes to an hour, will again enter the dream phase, stage 5 (REM), which this time is likely to last around twenty minutes. During the course of the night, the sleeper will repeat this cycle four or five times, each time staying in the REM phase for a longer and more vivid dream period. In this way, thousands upon thousands of dreams have been meticulously and laboriously monitored, giving researchers increasingly comprehensive insight into this puzzling, and evanescent phenomenon.

?

How do they decide whose obituary gets published in *The New York Times*?

"ALL THE NEWS that's fit to print," boasts the motto of the venerable New York paper, and the editors there apply the sentiment not only to foreign and domestic bureau reports but also to death notices. An obituary in the *Times* is the ultimate tribute: it signifies that during his life the deceased made it big—big enough that *The New York Times* considers his death to be newsworthy.

According to the *Times*, journalistic value is the only criterion for determining who gets in. The subject of an obituary must have accomplished things that had some impact on the way the rest of us live. In general, these accomplishments arise from the fields of entertainment (famous actors), the arts (famous writers), sports (famous athletes), politics (famous former office holders), or business (famous captains of industry). Fame, of course, is a relative term: a concert pianist, for example, might be known by every classical musician in the country but not by the wider public. In the end it is the obituary editor who, like any other journalist, arbitrates between what is news and what isn't. Which is not to say that selection is a wholly arbitrary affair.

To prepare for a notable figure's death, the newspaper actually begins compiling information on the person's life while he is still among the quick. In many cases, the future obituary subject conducts interviews expressly for this reason, pragmatically tying up loose ends much in the way other people compose wills and buy cemetery plots.

The *Times* retains staff obituary writers whose job it is to keep abreast of deaths and write up the deeds of qualified recently deceased people. In the case of a particularly famous figure, the notice might be written by someone outside the obituary department, usually the reporter who covered the subject in life. When, say, a baseball great dies, the occasion is reported by a sports writer, who brings a special understanding to the subject.

Because not even *The New York Times* has the resources to track everyone's term on earth, the paper often accepts the help of outside sources, including the Associated Press wire service and families of dead persons, who send in announcements. Again, it is the job of the obituary editor to decide which of the many AP and private notification candidates are newsworthy.

?

How do they know there are more stars in the sky than grains of sand on an ocean beach?

IF BILLIONS OF PEOPLE devoted their lives solely to counting, they could actually count the numbers of stars and sand grains, because neither figure is infinite. The universe and its starry contents come to an end somewhere out there beyond our wildest imaginings. And although it may feel like the number of grains of sand on a beach is infinite, those too are limited. Using eyeball estimates, mathematical calculations, physical laws, and sophisticated photography, Dr. Neil D. Tyson, an astrophysicist at Princeton University, has come up with the numbers. Here's how he did it, and what they are.

Ever since Newton wrote the laws of gravity, astronomers have used the formula $M = \frac{av^2}{G\pi}$ (in which M is mass, v is velocity) to calculate the mass of planets, stars, and even whole galaxies based on how fast these bodies travel in orbit. Astronomers know, for example, that the sun's mass is 2×10^{30} kilograms and that our whole galaxy, the Milky Way, has a mass of about 100 billion times that of the sun. Most of this mass is composed of stars similar to the sun, so we can safely say that our galaxy has about 100 billion stars.

If we take our galaxy as average, then with the number of stars per galaxy in hand, Tyson now needs to know the number of galaxies in the whole universe. A sophisticated photographic technique that uses a very sensitive digital detector, called a charge-coupled device, enables astronomers to take a very deep photo of the sky. Correcting for the fact that some galaxies are too dim to see, and estimating the number of galaxies astronomers would find if they shot photographs of every piece of the sky, astronomers have determined that there are about 10 billion existing galaxies. Tyson multiplies the 10 billion galaxies by the average of 100 billion stars in each galaxy to arrive at an estimate of the total number of stars in the sky: one sextillion, or a one followed by twenty-one zeroes.

Now on to sand. Tyson took a day trip to New York's Jones Beach and counted the number of grains of sand along a length of one centimeter. He counted about twenty-five. Allowing for some beaches to have much finer grained sand, as on the American Gulf Coast, and others to have big grains, Tyson estimated that there are about fifteen thousand grains of sand in a volume of one cubic centimeter on any given beach. He then judged that the average beach is about five kilometers long, one kilometer wide (from the edge to the high-tide line) and 0.015 kilometer (fifteen meters) deep. The volume of the average beach is therefore 0.075 cubic kilometers, or 75 trillion cubic centimeters. With 75 trillion cubic centimeters of sand on a beach that size, and fifteen thousand grains of sand in each cubic centimeter, the total number of sand grains on a beach would be about one quintillion, or a one followed by eighteen zeroes.

A sextillion is a thousand times bigger than a quintillion, so there are quite a lot more stars in the sky than grains of sand on a beach. Now let's say there are a thousand big beaches on planet earth; that would mean a sextillion grains of sand, or about the same number as all the stars in the universe. But then keep going. You could also add the sand grains under the ocean, and the sand on the Sahara Desert. "And that's the ball game," says Tyson. Sand grains win. By how much? No one's counted yet.

?

How do they decide where to put a new book in the Library of Congress?

IN 1814 Thomas Jefferson, retired from government service and living at Monticello, donated his personal library to Congress, writing, "There is, in fact, no subject to which a Member of Congress may not have occasion to refer." The government took Jefferson at his word, and the Library of Congress has since become the world's most avid collector of books, manuscripts, documents, music and film. It holds about 100 million items, and adds some seven thousand more every day. No one, not even the Librarian of Congress himself, has ever really examined the entire collection. Library officials claim that "if you were to limit your examination to one minute per item, eight hours a day, five days a week, it would take 650 years to view everything in the Library of Congress, during which time another 650 million items would be added behind your back." The massive holdings are stored on shelves that stretch for 575 miles hither and thither through the Library's three Washington, D.C., buildings, a storage facility in Landover, Maryland, and a film archive in Dayton, Ohio.

With items pouring in at the rate of ten per minute, the Library has by necessity become a leading innovator in the distinctly bookish fields of classification and storage. The Library classifies books according to the Library of Congress System, which has a broader base and greater flexibility than the Dewey Decimal Classification System. The system uses twenty-three letters of the alphabet to represent twenty-one general categories; combinations of letters are used to indicate specific topics. A, for example, stands for General Works, and AE designates encyclopedias. Subjects are further delineated by the use of three-digit codes after the letter combinations. There are full-time employees whose job it is to place new acquisitions on the shelves according to their classification. New shelves are constantly being added. Occasionally, when old books begin to decay and are transferred to microfilm, or when books are found to contain gross inaccuracies and are removed from the collection, additional shelf space becomes available.

Some items are not placed in the stacks at all. Newspapers are recorded on microfilm within three months of being published and are stored in microfilm drawers. Telephone books are treated the same way. Manuscripts, of which the Library receives more than half a million a year, both published and unpublished, are indexed and stored in manuscript boxes. The eight hundred or so motion pictures that arrive annually are placed in film canisters, some of which are stored in Washington, some in Dayton. Maps, numbering close to four million, are kept in specially designed flat drawers in the Cartography Division, which also houses an extensive collection of globes.

The Library has by law received two free copies of every book, map, chart, print, woodcut, engraving, dramatic or musical composition, or photograph submitted for copyright in the United States since 1870. The Copyright Office currently registers about 600,000 claims a year, and about half of its holdings are delivered to the Library. Copyright items not selected for the permanent collection are kept in the Landover, Maryland, warehouse. Also stored in Landover are Thomas Jefferson's fragile, two-hundred-year-old library and bound newspapers dating back to the seventeenth century. To help preserve its collection of antiquities, the Library has developed a patented diethyl zinc process for removing acid from book paper, a treatment that can extend the life of a book five centuries.

Another way to preserve printed material is to transfer it onto optical laser disks, which last forever. A single twelve-inch disk can store ten thousand printed pages, making it a valuable space saver as well. The library is currently evaluating the practicality of switching to laser disks.

In addition to its general collection, the Library maintains special branches that oversee specific fields of interest. The Hispanic Division, for instance, deals solely with items relating to Spain and Latin America, while the Asian Division counts among its holdings passages from a Buddhist sutra printed in A.D 770, the oldest example of such printing in the world. Certain books and materials defy usual classification and become the property of the Rare Book and Special Collections Division. Among the titles maintained in the special collections is *Ant*, which at 1.4 millimeters square is the size of an ant's book and the Library's smallest volume.

?

How do they discover a new drug?

IN THE EARLY HISTORY of mankind, before the advent of microwave dinners and Food and Drug Administration product labeling requirements, hungry men and women had to work hard to find a nutritious meal. Conceivably, to learn which plants and animals were edible and which were to be avoided, food pioneers scoured the earth like teething babies, popping whatever they found into their mouths. Rocks are no good, bark is no good, carrots are okay. Modern pharmaceutical development proceeds in much the same manner.

One place the search for effective disease-combating molecules begins is in the world's sludge heaps, where drug manufacturers, hoping to unearth microbes that contain new, medicinal molecules, go to collect dirt samples. Having amassed as many as fifty thousand compounds from various global dirt piles, drug company scientists begin randomly screening the compounds for useful attributes. The dirt compounds are cooked into a sort of laboratory witch's brew and tested against cells and viruses for activity. An "active" compound is one that modifies a healthy cell or kills a diseased cell or virus. After isolating an active compound, scientists must then identify the molecule behind its activity. If they are successful, and if the chemical agent is patentable, a new drug might show up on the shelves of the local pharmacy.

Drug screening's biggest drawback is its randomness. Unless the right compound is matched with the right disease, a hit-or-miss proposition, no new drug will be found. If an active compound is found, one that produces results against cancer, for instance, chances are pretty good that the effective molecule will contain superfluous, but inseparable, toxic elements. The molecule's toxic elements often cause side effects that seriously limit therapeutic potential. When scientists are able to isolate the active molecule within a randomly selected compound, and when the molecule's inherent toxicity is not self-defeating, they have the makings of a new drug.

Screening, as the scouring technique is called, has been the drug industry norm for fifty years, and given the odds, its success has been remarkable. One of the biggest achievements recently was the development of cyclosporine, a drug discovered in a dirt sample taken from just below the Arctic Circle in Norway. Cyclosporine prevents the immune system from rejecting transplanted organs. Belonging to the family of drugs called immunosuppressants, it has also shown promise in the fight against such diseases as multiple sclerosis, in which the immune system rebels against the body's native cells.

?

How do they get honey out of a honeycomb?

NO, NOT BY MACHINE ALONE. The beekeeper himself barehanded shoos away the swarming bees to carry off the honey-laden combs.

A man-made beehive—which will house a colony of fifteen thousand to sixty thousand bees—is constructed of stacks of wooden boxes, measuring twenty inches by sixteen and one quarter inches each. The uppermost boxes, called supers, will hold the surplus honey the hive produces—the harvest designated by the beekeeper for human consumption. Each box contains six or eight hanging frames, which hold sheets of pitted wax called comb foundation. During spring and early summer, worker bees in the hive construct wax honeycombs on the foundation using fat secreted from their glands. The comb cells are then filled with honey, which hive bees produce from flower nectar. By mid-July, after the clover and alfalfa have bloomed, the hives become so full of honey that a typical super weighs about sixty pounds. Then the bees start capping over the comb cells with wax.

How can the beekeeper get at the honey? First, he must get the bees to leave their combs. He wears a hat and veil to protect his face from stings during this procedure, but he usually doesn't wear gloves or other protective gear. The experienced beekeeper is accustomed to being stung, and in fact his body is resistant to swelling. Quite often he can avoid being stung at all. He knows which time of day to go about the business—sunny hours are better than chill and shadowy periods. On rainy, windy days, he can count on getting stung. Judging from the weather, then, and what the bees have had for forage, he can predict the mood of the swarms. Even if they're not irritable, he takes precautions to keep them calm. With a smoker, a bellows mounted on a little wood-burning stove the size of a large bean can, he settles down the hive with long puffs of smoke. Then he brushes the bees off the frames or allows them to pass through a one-way "bee escape" into the brood nest in the lower boxes of the hive. Some beekeepers use repellents to get the bees moving; some use forced air to blow them off the combs.

Now the beekeeper takes the supers to the extracting house, or another indoor spot away from potential robber bees. With a long, sharp, heated uncapping knife he shaves off the wax cappings. This is the best and lightest beeswax, which, after the honey drips off of it, can be sold for candle making. He suspends two to four frames inside a piece of equipment called a honey extractor: a centrifuge the size of a wide trash barrel. Large honey producers use bigger extractors that can accommodate several dozen frames. The extractor spins the frames inside it—it's rather like a washing machine on spin cycle. Centrifugal force sends the honey flying out from the combs and onto the sides of the tank; the liquid runs down and out a spout. The beekeeper collects it in a storage tank and strains it. Before being bottled, the honey is usually pasteurized by heating it.

How long does this whole process take? It depends on equipment and the size of the apiary, but usually a beekeeper allots about a day to pry the hives apart and get the supers inside. After that he can spread the extraction chores over a few days. How much honey can be harvested? A commercial beekeeper, with the best agricultural land and the best floral sources for his bees, needs to extract about a hundred pounds of honey from each colony per year. A hobbyist can usually be satisfied with sixty pounds.

After extracting the honey, the beekeeper stores the supers for next spring or, if he is hoping for a second harvest, he stacks the boxes on the hives again. When the bees are allowed back in, they begin repairing any damage the centrifuge did to their honeycombs.

?

How do they know what caste a resident of Bombay belongs to?

A CASTE, from the Portuguese word *casta*, meaning race, is a group of people sharing racial characteristics; in India the term is applied to the smallest endogamous groups existing within the general population. There are thousands and thousands of castes and subcastes in India, as many as thirty thousand by some estimates, which would make it very difficult for a resident of Bombay to identify, without asking, the specific caste of a fellow citizen. Traditionally, however, Indian castes have been grouped into four *varnas,* or classes, which is what most people mean when they talk about castes. In addition to the four varnas there is an outcaste group, whose members are at the bottom of the social hierarchy. The varna to which an individual belongs can be deduced from such things as that person's mode of dress, his trade, and his religious practices. The four classes are the Brahmans (priests), the Kshatriyas (warriors or barons), the Vaisyas (merchants), and the Sudras (laborers). The outcastes are called the Harijans, or "children of god," Mohandas Gandhi's name for the group once known as "untouchables."

According to ancient custom, the various jobs, dietary habits, and living conditions of each caste cause its members to become ritually unclean in varying degrees, with the highest caste, the Brahmans, remaining most pure, and the outcastes, the Harijans, being the most impure. For this reason, it was considered defiling to mix with people of lower social standing.

The caste system has been relaxed in modern times, and since 1947, when India gained independence, discrimination against "untouchables" has been illegal. These reforms are especially strong in great urban centers like Bombay, where it would be impossible for residents to completely avoid intercaste contact. Even so, the caste system continues to exert a strong social and political influence in India.

In modern, industrial Bombay, recognizing what caste another person belongs to is basically a matter of cultural literacy: each person advertises his social position by his name, the clothes he wears, the job he performs, and the food he eats. A Brahman, for example, will likely adhere to a strict vegetarian diet for religious reasons. A member of one of the lower classes will be less particular about what he eats. A Harijan will probably eat whatever food he can obtain, regardless of its ritual impurity. Some distinctions run along fairly universal class lines: a person who lives on the streets or in a slum is apt to belong to a lowly caste, while a person who resides in an elegant apartment in one of Bombay's exclusive neighborhoods probably belongs to a favored caste. Clothing tells a similar story. The quality, material, and style of a person's clothing, whether traditional or Westernized, says something about his social standing. There are hundreds of clues, and a native of Bombay would be proficient at reading them.

This situation is much the same in any culture. In the United States, for example, the name Rockefeller carries a certain cachet. If you were to meet a man by that name, and he happened to be getting out of a limousine wearing a very expensive-looking suit and top coat, you might assume a thing or two about his background.

The great difference between Indian castes and American classes is that the Indian system permits only very limited social mobility. In Bombay, members of different castes by necessity mingle during the course of a day, but they rarely become involved permanently. Marriage especially is governed by social taboos. While the sons and daughters of American Rockefellers might meet and marry daughters and sons of less socially prominent families, that is not likely to happen in India.

In India, a caste near the bottom of the social ladder might advance itself by adopting the customs of the caste above it. After a few generations, it would become almost indistinguishable from the higher caste, but there is no real upward movement of individuals.

How do they dry-clean clothes without getting them wet?

THE TRUTH IS DRY CLEANING is not entirely dry. Dry cleaners do apply liquids to silk dresses . . . it's just that the stuff contains no water and furthermore dries quite rapidly. It doesn't cause clothes to shrink or colors to run.

Routine dry cleaning is done in large, closed machines using a volatile chemical called perchloroethylene (C_2Cl_4), dubbed "perk" in the trade. Perk is a compound consisting mostly of chlorine, with a lesser part of carbon. As clothes are agitated in a machine, perk, along with a detergent, passes through the machine at a rate of two thousand gallons an hour. About every minute there's a new batch of perk running through the clothes, dissolving stains and leaving dirt stranded in a filter. After this process most clothes are ready to be pressed and bagged.

The meticulous dry cleaner may, however, spy a stain that withstood the perk and need to apply the finer skills of his trade. The first step—a crucial one—is to analyze and identify the stain. Many are unknown; a large proportion, called "sweet stains," contain sugar, salt or starch and come from spilled food. The spotter then selects from four standard methods of stain removal and goes to work at his spotting board. He may dissolve the stain with the appropriate solvent, such as volatile-type paint remover, hydrocarbons, or petroleum distillates. (Legend has it that the dry-cleaning industry originated with a Frenchman in the nineteenth century who accidentally doused his clothes with benzene and subsequently found grease stains easy to get out.) If the stain is water soluble, the spotter can just blast it with his steam gun and blow it dry. Another approach is to apply a lubricant with a stiff brush or spatula and try to raise the stain to the surface of the fabric, where it can be removed more easily. If that tactic fails, the spotter may resort to chemical action: he applies an oxidizing bleach that removes the color from the stain and so camouflages it; or he converts the stain using acids or alkalis into a soluble substance that can be treated with solvents. The very last method is to digest the stain with enzymes, just as your gastrointestinal enzymes break down insoluble protein, albumin, and starches. The enzymes, which are from bacteria and yeast, convert insoluble substances into soluble ones that can be cleaned away quite easily.

The spotter, then, has an array of methods and chemicals at his fingertips, so you needn't despair when salad dressing drips down your tie. Just remember what you spilled, and half the battle is won.

?

How do they make glow-in-the-dark toys?

FROM DINOSAURS TO LEGO GHOSTS, from stickers to yo-yos, a battery of toys now add a cool greenish glow to their attraction.

The key ingredient that causes the glow is a phosphor (from the Greek for "light-bearer"), which has the ability to *phosphoresce:* it glows when exposed to visible light energy, and continues glowing for a period of minutes or hours after the exposure is stopped. This phenomenon occurs because the phosphor absorbs energy while exposed to light and later, in the dark, discharges the energy in the form of light. (A substance that *fluoresces,* on the other hand, stops emitting light once the source of energy is removed.)

In nature there is a chemical element called phosphorus, discovered about three hundred years ago by a German scientist, which has the remarkable ability to glow. Glow-in-the-dark yo-yos, however, contain a commercially made phosphor, often a compound of zinc or magnesium. The phosphor is made by heating to above 800 degrees Celsius pure, synthetic ingredients in the form of tiny crystals two to fifteen microns in diameter, along with an "activator," or trace of heavy metal. After being heated, the atoms of the activator enter the crystal lattice of the phosphor and remain embedded there.

When light strikes the phosphor crystals, the electrons that orbit around the activator's atoms are thrown into a tizzy. They leap from their normal, relaxed orbits into an excited orbit farther away. It is a rule of nature that everything wants to be in its lowest energy state, so the atoms continually plummet back toward the nucleus. Because of the lattice structure of the crystal, some electrons get trapped, their descent is delayed, and they continue to release energy—in the form of light—over time. Oddly enough, the smaller the crystal the longer the glow.

"A good glow-in-the-dark bug will shine for as long as thirty minutes after you shut off your lights," says Robert Majka, project manager at Canrad-Hanovia, Inc., a company in New Jersey that manufactures phosphorescent pigments. "Eighty-five percent of its luminescence will be expelled by then. But even after that, the bug will continue to glow for as long as eight hours, with the 15 percent of its leftover luminescence. You can only see this low glow if you've been in the room with it, so that your eyes have adjusted to the dark and the fading bug."

Canrad-Hanovia makes a glow-in-the-dark pigment containing an inorganic zinc sulfide phosphor, which is nontoxic and nonradioactive. It comes in the form of a powder, which is sold to compounders who add it to any clear suspension, be it plastic, paint, vinyl, or ink. A carrier material that is not clear would absorb the phosphorescence. From these plastics and inks come glow-in-the-dark Frisbees and mobiles, reptiles and stickers. The brightest toys with a glow that lasts the longest have a higher concentration of pigments than those with a dimmer glow. Optimal luminosity occurs when the pigment constitutes 30 percent of the product, that is, the ratio of plastic resin to pigment is about 2 to 1. At that point, the carrier reaches a saturation point.

Most kids know how to make their Super Balls and alligators glow. "Simply hold the toy up to any light for about thirty seconds," says Majka. Kids may be interested to learn that you get the brightest glow if you use fluorescent lights.

?

How do they predict when a fine wine will be ready to drink?

"AGE," WROTE FRANCIS BACON, "appears to be best in four things: old wood best to burn, old wine to drink, old friends to trust, old authors to read." A fine enough sentiment, but it turns out that most wines, including virtually all rosés and white table wines, peak in taste within a year of being bottled and after that begin a steady decline. Predicting when these wines will be ready to drink is like betting on the outcome of a fixed fight: it's the proverbial sure thing. But there is a relatively small number of wines, representing the high end of the market, that actually improve with age, and predicting a maturation date for them is neither as easy nor as precise.

Wine is nothing more than fermented grape juice. Unlike hard liquor, which is preserved by its high alcohol content, wine is perishable and loses freshness when it is exposed to air. Although some oxidation is good for some wines, increasing the wine's complexity of taste and aroma, too much oxygen will spoil it. Since wine continues to age in the bottle, where it is exposed to oxygen that either seeps through the cork or is present in the wine itself, the trick is letting it ripen to maturity but not beyond. A number of variables, including type of wine, variety of grape, vintage, and storage conditions, help experts guess when a wine's time has come.

Red wines, such as cabernet sauvignon, red Bordeaux, red Burgundy, and pinot noir, have a higher fixed acidity than white wines and tend to age better because of it. Based on their experience with these types of grapes, vintners are able to estimate when wines made from them will come to full maturity in the bottle. But the acidity of any one type of wine, cabernet sauvignon for instance, will vary from year to year and from vineyard to vineyard, depending on weather and soil conditions. In Europe, where the grape harvest is subject to quirky weather, winemakers apply the term vintage to years when conditions have been especially favorable to grape growing. Grapes from a vintage year, presumably, will have higher sugar and tannin contents than nonvintage grapes, and will therefore age better. In California, where the climate is very consistent, vintners are fond of saying that every year is a vintage year, but even there grapes and wines vary from year to year and the vintage label often signifies nothing more than a wine's age.

Experienced wine makers admit that there is no surefire way to predict when a wine will be ready to drink. Some vintages, heralded as masterpieces when they are bottled, rush headlong to maturity and become tired long before anyone expects. Others, like the 1794 Château Lafite, somehow survive their makers by more than a century.

Connoisseurs do know a few things that help them make well-informed guesses. Small bottles have a larger cross-section-of-neck to volume-of-bottle ratio than large bottles, and since exposure to air ages wine, and since air enters a bottle through its cork, a half-bottle of wine will mature before a full-size bottle of the same wine. Storage temperature is another important consideration. Wine is stored in cellars and not attics for the good reason that warmth accelerates the aging process. But too much cold will arrest the wine's development. A temperature of between 55 degrees and 60 degrees Fahrenheit is considered optimal. The single most important factor, though, in predicting wine maturity accurately is experience. Connoisseurs are able to predict when a wine will reach its peak of flavor by knowing its vineyard and the record of other wines from that vineyard. Based on a vineyard's past performances, wine experts have been able to draw up charts that show when a wine will in all likelihood mature. A chardonnay, for example, will improve for about three or four years, maintain its peak for a few more, and then decline. With a heavier, more tannic wine such as cabernet sauvignon, the general rule of thumb is six to fifteen years, depending, of course, on the growing and storage conditions.

Even the most astute student of wine will be caught off guard from time to time though, and some types of wine, such as claret, a deep red Bordeaux, simply defy precise forecasting. The best you can do with claret, experts advise, is wait ten to a hundred years before opening it, and maybe you will get lucky.

?

How do they know how much an aircraft carrier weighs?

HOISTING THE 100,000-TON *Nimitz* onto a scale is a formidable idea and not a very practical one. Fortunately, there are less cumbersome ways of determining the weight of such a massive ship.

If a new aircraft carrier is being built, technicians simply keep a running tab of all materials that are used in the construction. In some cases the weight of a particular piece may not be known but is estimated based on its size and material. The weight of electrical wiring and drainage piping, for instance, is difficult to calculate precisely and must be approximated. In addition to machinery and equipment, one must tabulate the ship's loads, consisting of fuel oil, water, ammunition, and a variety of stores. As older aircraft carriers are modernized they become heavier, owing to the installation of new equipment and support structure. Many of the airplanes carried become heavier, too. The *Enterprise,* for example, weighed a mere 86,800 tons in 1961 and now weighs closer to 92,800 tons. (The tons referred to here are long tons, each of which equals 2,240 pounds, as compared with a short ton, which is 2,000 pounds. The Navy uses long tons.)

Detailed weight records are maintained throughout the life of an aircraft carrier, but since some weights are estimated, the "paper ship" must be checked periodically against the real ship to determine its actual weight. To do this the Navy routinely reads the draft marks located on the sides of the carrier, at the bow and stern. The drafts are six-inch-high Roman numerals painted on the ship at one-foot intervals in a vertical line. These marks tell the Navy how much of the carrier's hull is underwater. By knowing the volume of the underwater hull, the Navy can figure the weight of the entire ship.

One can calculate the ship's weight because of Archimedes' principle, which states that a body immersed in fluid is buoyed by a force equal to the weight of the displaced fluid. This means that a body floating in water displaces a volume of water equal to its weight. To compute the weight, one multiplies the volume of water displaced by the density of water (weight = volume × density). To start, imagine a rectangular box twenty feet long, ten feet wide, and five feet deep. Let's say it rests in the water and sinks down two feet. The volume of the box that is submerged, which is also the volume of water displaced, is $20 \times 10 \times 2 = 400$ cubic feet. Now, the density of one cubic foot of sea water is 64.0 pounds, so the weight of the box may be calculated as follows: 400 cu. ft. × 64 lbs./cu. ft. = 25,600 lbs., or 11.43 long tons.

Typically, an aircraft carrier might have a draft reading of thirty-five feet, which means thirty-five feet of the hull is submerged. One multiplies thirty-five feet times the width and length of the ship to determine the volume that is submerged. Multiply this figure by the density of sea water and you have the weight.

Does it sound too simple? You're right. The tricky part is that the hull of the ship is not rectangular. "Although larger ships are somewhat boxy," says William Austin of the Naval Sea Systems Command, "all carriers are faired at the bow and the stern to allow them to move through the water more efficiently. From above, the outline of the hull would appear more like a tear shape than a box." Furthermore, as thc hull rises from the keel, it slopes outward so that the volume is constantly increasing. In order to calculate the volume of this three-dimensional, nonlinear shape, an engineer must make a number of area calculations at various intervals above the keel. As an example, consider five horizontal planes slicing through the hull. One must compute the area of each plane and multiply it by the distance between it and the next plane. Each of these smaller volumes is then added to the next volume to obtain the total volume up to a specified draft. Converting the incremental volume to weight and plotting it against the height above the keel produces a curve which can be used by the ship's personnel to determine the ship's weight each time the drafts are read. The procedure is lengthy and is done only once—at the time the ship is built. As long as the shape of the hull is not changed during the life of the ship, the curve developed to determine the ship's weight will remain accurate.

?

How do they decide what or who gets honored on U.S. postage stamps?

MAKING STAMPS IS A FUNCTION of the federal government, and as is the case with most governmental projects, a committee has been appointed to get the job done. The fourteen members of the Citizens' Advisory Stamp Committee, representing expertise in American art, business, history, and technology, and sharing an interest in philately, are handpicked by the postmaster general to recommend subjects for commemorative stamps. The committee convenes every two months to sift through the thousands of suggestions that pour in continuously from the provinces. Of the twenty thousand or so subjects submitted by the general public each year, twenty-five to thirty-five eventually make it through the committee to the postmaster general; once he gives final approval, they can become stamps.

During the watershed postal reorganization in the early 1970s, one of those events that go largely unnoticed by the public but retain deep significance within the government, the postmaster general, with help from the CASC, established the elementary criteria by which stamp subject selections are made. Since then the list has been expanded to include twelve points:

1. U.S. postage stamps and stationery will primarily feature American or American-related subjects.
2. No living person shall be portrayed on U.S. postage.
3. Stamps honoring individuals will be issued in conjunction

with anniversaries of their birth, but not sooner than ten years after the individual's death. U.S. presidents are the only exception to the ten-year rule, and may be honored on the first birth anniversary following death.

4. Events of historical significance shall be considered for commemoration only on anniversaries in multiples of fifty years.
5. Only events and themes of widespread national appeal will be considered.
6. No commercial enterprise; specific product; or for-profit fraternal, sectarian, political, service or charitable organizations shall be recognized.
7. Towns, cities, counties, municipalities, schools, hospitals, libraries, or similar institutions shall not be considered.
8. Postage stamps commemorating statehood anniversaries will be considered only at intervals of fifty years from the state's entry into the Union. Other anniversaries pertaining to individual states or regions will be considered only for postal stationery and only at intervals of fifty years from the date of the significant event.
9. Stamps shall not be issued to honor religious institutions or individuals whose principal achievements are associated with religious undertakings or beliefs.
10. No "semipostals," stamps to be sold at a premium over their postal value to raise money for charitable organizations, shall be issued.
11. Significant anniversaries of universities and other institutions of higher learning shall be considered only in regard to Historic Preservation Series postal cards featuring an appropriate campus building.
12. No subject, except traditional themes such as Christmas, the U.S. flag, Express Mail, Love, and so forth, will be honored more than once every ten years.

Anyone can petition the Citizens' Advisory Stamp Committee with an idea, and if the proposed subject meets the guidelines, it will be considered and possibly recommended to the postmaster general. The rules, in addition to telling when and how something can be commemorated, dictate to some degree the subject matter itself. A stamp might be issued, for instance, to celebrate food in general, but rule number seven will prevent a specific product, like Spam, from ever gracing U.S. postage. Likewise, the government will not issue a stamp to honor one particular hospital, but health care in the abstract could be saluted.

?

How do emperor penguins stay warm in Antarctica?

THE EMPEROR PENGUIN, which happily swims and romps in killing temperatures of minus 40 degrees Fahrenheit, is the only bird that can, and often does, spend its entire life on coastal or shelf ice, without ever coming ashore. Basically, three mechanisms are at work in safeguarding the emperor penguin's body heat: its feather construction, its weight, and its behavior.

The emperor penguin's body is totally covered with feathers; it has no bare patches around the feet or head as are found on penguins of warmer climates. The tiny, overlapping feathers, numbering about seventy per square inch, are packed so closely they form an impermeable barrier to the wind. Coated with oil, they also resist water. The shafts of the feathers have tufts that, along with a layer of air next to the skin, provide insulation. Below the skin is a layer of blubber thicker and heavier than that of birds in milder climates.

Penguins in Antarctica are in general bigger and heavier than other penguins, and it serves them well. As George Gaylord Simpson explains in his book *Penguins,* heat loss is proportional

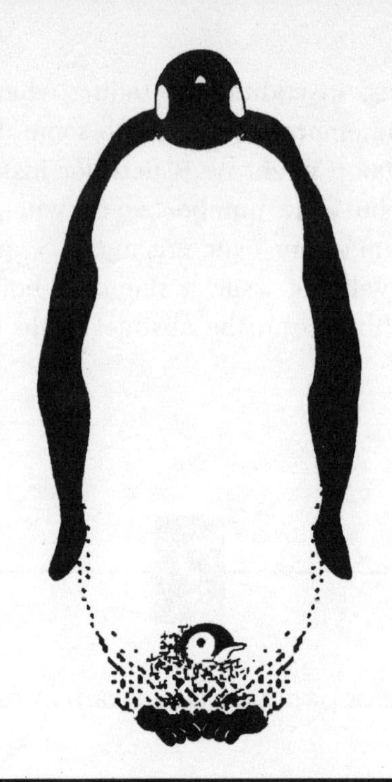

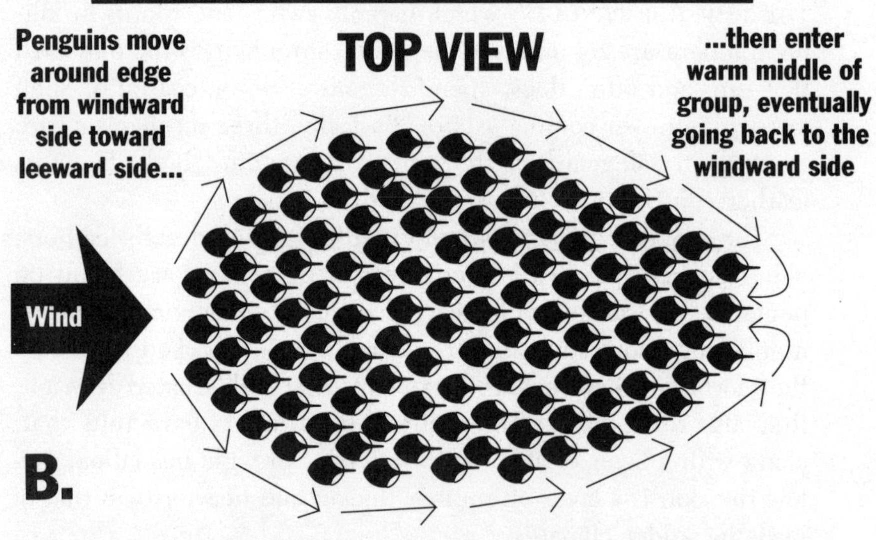

A: A baby emperor penguin gets off the ice by perching on its parent's feet. It nestles against the soft belly feathers for protection against the piercing wind.
B: Emperor penguins huddle in groups for warmth. Individuals systematically move from colder to warmer spots in the group, thereby taking turns in the warm center and ensuring the group's survival.

to the ratio of surface area to body volume. Larger animals have a lower ratio of surface to volume, and consequently proportionally less surface area from which heat escapes. The emperor's average weight is sixty-six pounds, as compared with the Galápagos penguin, which at five pounds enjoys balmy temperatures and seventy degree water.

Perhaps the most incredible survival feat of the emperor penguin is its ability to raise young in inhospitable Antarctic weather. The female lays one egg, usually in May, and then leaves the homefront, sometimes ranging as far as sixty miles in search of food. Papa is left to guard the egg. He holds the precious cargo on his feet, protected under a fold of skin, and does not leave his post for sixty-four days. He eats nothing during this period. At about the time the young hatch, the females return and the weary males can at last go in search of food. The chicks have only fluff at first, not a full suit of feathers, so they need a shield against the cold. Perched on the parent's feet, beneath layers of feathers, the baby snuggles against the grown-up's belly.

The adult emperor penguins, meanwhile, huddle against each other in a large circular mass to maintain body heat. They turn their backs to the piercing wind, and can raise the temperature within the group by as much as 20 degrees Fahrenheit. Those positioned farthest upwind, that is, those with their backs most exposed, periodically move around the outside of the huddle to the front of the group. As more penguins from the back move up, those at the front wind up in the middle—the warmest spot. The well-coordinated system continues so that each penguin in turn shifts from colder to warmer to colder spots in the huddle.

In these ways emperor penguins manage to keep quite warm in their Antarctic habitat. So warm, in fact, that penguin expert Bernard Stonehouse feels the real problem is: How can the emperor penguin ever cool down? Given its impervious feathers and blubber, how can it let off heat when intensely active? The emperor penguin's solution is to ruffle its feathers, allowing body heat to escape, and spread its wings, exposing more of its body to the icy Antarctic winds.

?

How do they make it easier for celebrities to attend concerts or Broadway shows?

IN SHOW BUSINESS, getting good seats at a performance is simply a matter of knowing the right people. And celebrities tend to know most of them.

Before tickets go on sale for any concert or Broadway show, a certain percentage are put aside as "comps," or complimentary tickets. Say, for example, Joe Rockstar plays a gig at the XYZ Theater and has signed a contract with a promoter to publicize the event. The contract includes a provision that sets aside a certain number of free tickets for the show. These tickets are divided among the performers, the promoter and the management of the building in which the show will take place. So if 1 percent of the seats in a ten-thousand-seat venue are designated comps, Joe Rockstar might get sixty, while the promoter and the XYZ Theater pocket twenty apiece. The breakdown varies according to the particular contract.

Now let's say Joe Rockstar wants to go to a Marilyn Operastar performance that is already sold out. If he knows Marilyn, then it's easy—he calls her and she gives him one of her comps. If he doesn't know her, it's still pretty easy. Chances are Joe or his manager knows Marilyn's manager, or her promoter, or the management of the theater in which she will perform. One of them supplies some comps. In the event that Joe doesn't know anyone who knows anyone, he could as a celebrity approach the performer, the promoter, or the theater, introduce himself and request a ticket. If Joe is indeed famous, he'll probably get a comp as a professional courtesy if one is available. The promoter or the theater might in fact have some professional interest in doing Joe Rockstar a favor . . . one or both of them might want to put on a Joe Rockstar concert one day. That's show business.

?

How do they treat drug addiction with acupuncture?

IN TREATING DRUG ADDICTION, acupuncture is used to suppress the addict's physical craving for narcotics and also to combat such debilitating side effects of withdrawal as nausea and extreme anxiety. By easing an addict's symptoms of addiction and withdrawal, acupuncture facilitates the patient's ability to receive other therapies, including counseling.

Acupuncture treatment is based on the presence in the body of certain pathways through which a vital energy, called *qi* in Chinese, flows like a bioelectric river. The pathways, or meridians, are each associated with a specific bodily function and interconnect with one another to form one continuous loop. Ancient Chinese physicians hypothesized that a blockage of *qi* along its meridian caused the base organ to malfunction, resulting in disease. They charted the meridians and discovered more than 365 points on the body at which *qi* could be reached and stimulated by inserting needles into the body. Because of the intricacy of the meridian system, the acupuncture points associated with an organ are often far removed from that organ. The stomach meridian, for instance, begins under the eye and descends all the way to the toes. The thirty-sixth point of the stomach meridian, located below the kneecap, is used to treat digestive disorders such as gastritis and dyspepsia. Modern tests show that the acupuncture points possess an electric current distinct from that of the rest of the body.

The acupuncture points commonly used in treating drug addiction are auricular, with five on each ear. Stimulation of these points with thin flexible needles causes the body to release organic opiate-like chemicals that defuse the craving mechanism, enhancing an addict's mental and physical stability. When properly administered, drug treatment acupuncture produces a sharp increase of metenkephalin in the subject's cerebrospinal fluid. Metenkephalin is a chemical with analgesic properties that occurs naturally in the human brain. A standard acupuncture detox program lasts for two weeks and consists of twelve forty-five-minute-long sessions. All ten of the patient's ear points are stimulated simultaneously, which means that during a session he will have ten needles protruding from his head. While this picture may seem unpleasant to most of us, urban clinicians note the fortunate circumstance that intravenous drug users seem to be undaunted by all the needles.

Detoxification is a relatively new application of the ancient medicine. The Chinese, looking for a way to curb opium abuse, first experimented with acupuncture as an antidote to addiction in the 1940s. In this country, acupuncture detox was developed in the 1970s, at Lincoln Hospital in the Bronx. The Lincoln program, which combines acupuncture and intensive A.A.-type counseling, treats 250 addicts a day and maintains a success rate of 50 percent. At about twenty-five dollars per person per visit, acupuncture costs far less than any inpatient rehab program, a factor that has led many cities around the country to adopt the Lincoln model.

In 1989, a group of doctors from Minnesota published the results of a study they had conducted on the efficacy of acupuncture treatment for chronic alcoholism. Eighty severe, long-term alcoholics participated in the study. Of these, half belonged to a treatment group that received acupuncture at points specific for substance abuse. The rest, who made up the control group, received acupuncture at nonspecific points on their ears. The study was divided into three phases. During the first phase patients received five acupuncture treatments a week over a two-week period. Frequency of treatment was reduced to three times a week and twice a week in the second and third phases, respectively. The patients were then discharged and asked to return to the hospital after one, three, and six months had elapsed.

The doctors in charge of the study reported that 52.5 percent of the patients in the treatment group, as opposed to only 2.5 percent of the control-group patients, completed all three phases of the program. During the follow-up period twice as many control patients as treatment patients reported episodes of drinking. Also during that time, twice as many control patients as treatment patients were admitted to a detox center. Of the twenty-one patients who were admitted to detox at least five times over the six-month study period, fifteen came from the control group, five were treatment patients who had dropped out of treatment early, and only one was a treatment patient who had completed all three phases of the program.

Although acupuncture cannot "cure" drug addiction, studies like the one in Minnesota show that it can be a clinically effective treatment.

?

How do tightrope walkers stay on the wire?

THERE ARE NO REAL TRICKS to staying on a tightrope. One does it the same way one gets to Carnegie Hall: practice, practice, practice.

Angel Quiros, a high-wire artist for Ringling Brothers and Barnum & Bailey Circus has been walking the wire for sixteen years. One must be in good physical condition, he says, "but the main thing is to practice a lot." Acrophobes need not apply.

Quiros learned his trade in his grandfather's circus in Madrid at twelve years of age. First he and his siblings tried their natural abilities on a taut elevator cable held three feet in the air. They got up, tried to walk and often fell the short distance to the ground.

Within three years they were walking, running, dancing, jumping rope, doing gymnastics and sword fighting on the cable, still at an altitude of three feet. When they had mastered these skills, they gradually raised the cable in approximately seven-foot increments until the Quiros clan was doing its act some forty feet up in the air. This is the height at which Quiros performs for Ringling Brothers with no net below and no safety wire—a clear line attached to the artist's belt.

Using a pole gives the high-wire artist a better sense of balance—when Quiros uses one he is able to carry his brother on his shoulders—but the added weight of the pole slows down running and makes jumping more difficult. It's also easier to balance barefoot, but this is tough on the feet, so Quiros wears Capezio dance shoes with rubber soles.

Still, no matter how much an aspiring tightrope walker practices, there is no guarantee that he'll stay on the wire. Angel Quiros has had broken ribs to prove it.

?

How do they recycle newspapers?

FOR YEARS, mountains of old newsprint were blithely tossed out along with other household and commercial waste. Today, as we slowly realize our natural resources are less than infinite, we are recycling about a third of the country's newspapers. In order to reuse paper, one must first return it to the consistency of virgin wood pulp. After that, it can be cleaned and prepared for a second (or third or fourth) feeding through paper-making machines. Let's start at the beginning of a typical mill process.

First the newspapers are collected by towns or commercial trash haulers and separated from higher grades of paper (the latter, incidentally, can be turned into higher-grade final products). Then the newspapers are either bagged in brown paper or bundled with string, sold to mills and stored in warehouses until use. When a mill begins to fill an order for recycled paper, its conveyor belts dump measured amounts of the newspapers, with water, into a giant thirty-foot-diameter blender called a hydrapulper. Rotating blades loosen and roughen the cellulose wood fibers; reduced to a slushy, papier-mâché-like consistency, the newspapers are now called a "slurry."

To remove staples and paper clips, the slurry is now fed into rotating cylinders with coarse screens that trap the contaminants while letting the slurry escape through thousands of holes. Fine screens remove the smaller impurities, particularly the small pieces of binding glue called "stickies" that can make several sheets of paper tear in the presses later on. The mixture is now called "stock."

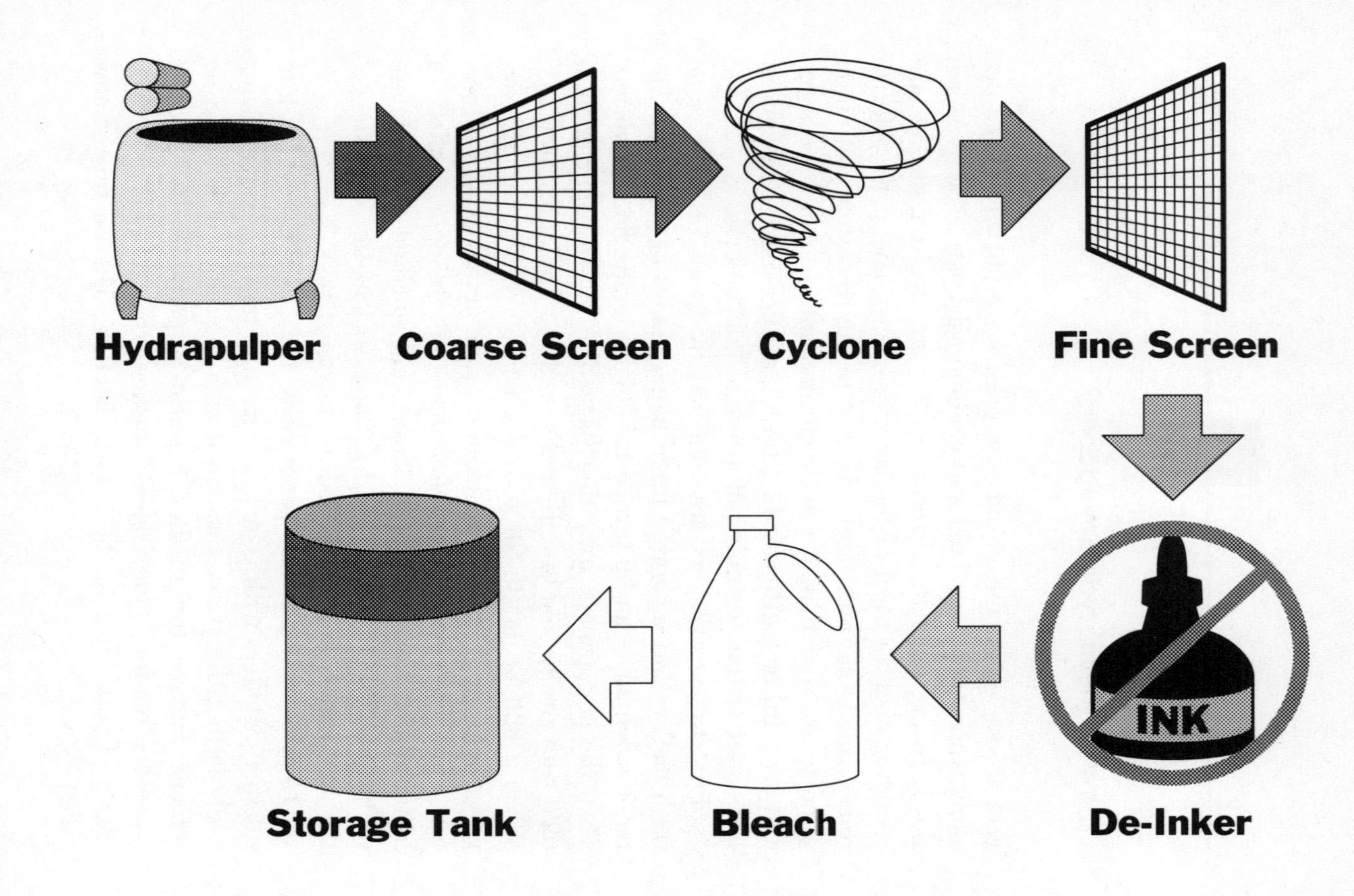
Hydrapulper
Coarse Screen
Cyclone
Fine Screen
INK
De-Inker
Bleach
Storage Tank

In newspaper recycling, first the newsprint has to be reduced to a pulp, then it can be rebuilt into paper. Here you can see both halves of that process: In the steps from hydrapulper to storage tank, the newspaper becomes clean, deinked pulp.

BELOW: From header box to the finished roll, the pulp is shaped, dried, and smoothed into paper again.

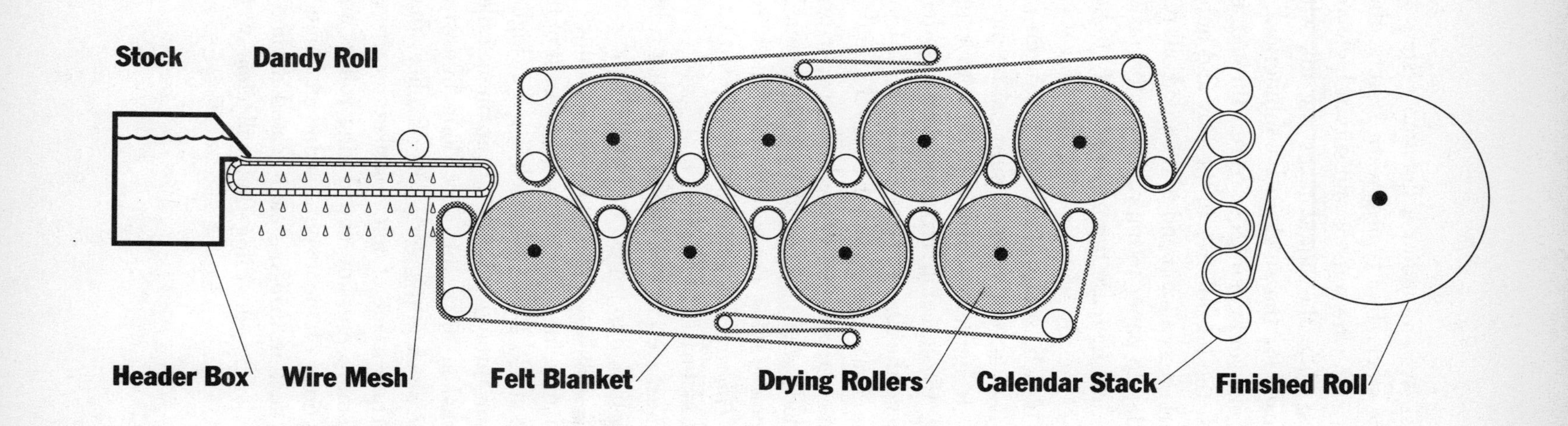

A wash system and/or a flotation system is used to strip the ink off the stock. In a typical wash system, the stock is shot with liquid through wire screens and slits at high speed; the wires grab the fibers while the liquid takes the ink with it. The flotation method calls for the stock to be submerged in a tank while air bubbles, clay and chemicals are run through it. Under high temperatures, the ink adheres to the clay and bubbles; it gets carried to the surface as a foam and is skimmed off. (Clay, incidentally, is often added to paper to improve opacity, make the paper surface smoother and increase the paper's ability to hold ink.) If an order calls for especially white paper, the mill washes the stock with chlorine bleach, sulfur dioxide or hydrogen peroxide.

At this point virgin wood pulp can be piped into the recycled stock to strengthen it. It's important to keep in mind that the slurrying, the forceful screening, and the de-inking will eventually damage the fibers in any paper, so that unlike aluminum and glass, the same paper can't be recycled indefinitely. Pure recycled paper can clog high-speed newspaper presses: adding virgin pulp reduces the chances of that happening. Barry Commoner, in his book *Making Peace with the Planet*, reports that *The Los Angeles Times* prints its news on paper that's about 80 percent recycled material.

The next step in newspaper recycling is building the fibers in the stock back into paper. In a massive piece of equipment called, rather predictably, a paper-making machine—some are as long as two football fields—more than 90 percent of the water is removed from the stock. First, the stock is fed at high speed in a continuous stream through a slit in a component called the head box and gently shaken on a thirty-foot-long fine-wire mesh. The mesh feeds the stock into several sets of rollers that squeeze out more water. Then a set of felt "blankets" carries the still-wet stock through more pressing and drying rollers and delivers it to the final sections of the paper-making machine. These can include vacuum suction driers, steam-heated drying drums and banks of infrared lamps—anything that will clear water from the paper.

Lastly, the paper stock is peeled out of the end of the machine and wound on rolls for delivery to the printer.

Recycled newspaper can't be made into fine stationery, but it can effectively be turned into such products as brown paper, toilet paper, cereal boxes, construction paper, egg cartons, insulation, animal bedding, cat litter, soil enhancer, telephone directories and, of course, newsprint.

?

How does a magician pull a rabbit out of a hat?

THE WHITE RABBIT and the black top hat have long been symbols of the magician's art—his uncanny ability to conjure objects and cause them to vanish again. Rabbits first appeared onstage, popping out of top hats, in the late 1830s. Though the originator of the trick is unknown, John Henry Anderson of Scotland, dubbed The Wizard of the North, was among the first to perform it, in 1840, at the New Strand Theater in London. Another nineteenth-century conjurer, Joseph Hartz, caused a skull to rise out of a top hat, while the French wizard Jean-Eugène Robert-Houdin produced nothing less than a cannonball.

Whatever the object, the technique remains virtually the same. The magician needs a table, draped with a tablecloth, which prevents the audience from seeing a small shelf, known as a servante, perched at the back of the table. Sometimes the tablecloth itself is simply pinned up, thereby creating a small pocket in which to house the rabbit or other object to be produced. The magician wraps a large silk handkerchief around the bunny and

In this classic trick, the magician first conceals the rabbit in a handkerchief on a small shelf behind the table. He then grasps the bundle along with the brim of the hat and, while flipping over the hat, swiftly moves the rabbit bundle inside. After a wave of the wand, the magician pulls out the rabbit.

ties the ends with a rubber band. At the outset of the trick, he removes his hat and displays the inside—empty. He may even pass it around the audience for all to inspect. When it has been returned he sets it down, right side up, near the back of the table. While waving his wand with his right hand, he grasps both the brim of the hat and the corners of the handkerchief with his left. (The ends of the handkerchief are sandwiched between his thumb and the brim and therefore well concealed.) With a swift, graceful move, he turns over the hat. The bundle drops into the hat, he removes the rubber band, and presto, with another wave of his wand he raises the live and wriggling rabbit into the air.

?

How do they make long-life milk?

MOST OF US WOULDN'T consider drinking milk that has sat on the supermarket shelf *unrefrigerated* for some three months. Then again, many people, especially those without small children, have never heard of ultrahigh-temperature (UHT) milk.

Farm Best milk, made by Dairymen in Savannah, Georgia, is one of a number of long-shelf-life milks now available. You'll find it not in the dairy section but probably along with boxes or cans of juice that are not refrigerated. How can it sit there so long without spoiling?

First, this milk is processed at 280 degrees Fahrenheit rather than at 170 degrees Fahrenheit, the normal temperature for pasteurizing milk. The high temperatures kill microorganisms that might escape unscathed by conventional processing. What you have, in fact, is sterile milk, totally free of any spores or bacteria that could multiply and cause the milk to spoil.

Secondly, the product is securely packaged to keep out air, light, and bacteria. After the milk has been purified by the UHT process, it is transferred to an aseptic packing machine which loads the milk into sterilized cartons. These sturdy containers are made up of five layers of materials—three polyethylene, one paper, and one aluminum foil—and hermetically sealed. You can keep the cartons for months. Once you open them, you should refrigerate the milk, but even then it won't turn sour for about three weeks.

?

How do they pick Pulitzer Prize winners?

PULITZER PRIZES ARE AWARDED annually for distinguished work in twenty-two categories of journalism, arts and letters, and music. The fourteen prizes in journalism are meritorious public service, spot news reporting, beat reporting, national reporting, international reporting, investigative reporting, explanatory journalism, editorials, editorial cartoons, spot news photography, feature photography, commentary, criticism, and feature writing. The six arts and letters categories are fiction, drama, general nonfiction, history, biography, and poetry. One prize is awarded in the music category, for a "distinguished musical composition . . . which had its debut in the United States during the year."

The meritorious public service prize is given to an outstanding newspaper; all other prizes go to individuals who have distinguished themselves in their various fields during the year. Unlike Nobel Prizes in literature, which are awarded to writers on the basis of their whole body of work, Pulitzers recognize specific works. John Steinbeck, for example, won the Pulitzer Prize for fiction in 1940, for *The Grapes of Wrath;* in 1962 he won a Nobel Prize in literature for his achievements over the course of a lifetime.

The prizes are administered by Columbia University, which relies on the Pulitzer Prize Board, an eighteen-member body composed of publishing executives and academicians, to select the prize recipients. Because of the huge volume of submissions, the board depends on prize juries to screen the entries. Members of the juries—there is one jury for each prize category—sift through all the entries in their category until they come up with three finalists. Each jury submits its list of finalists to special three-member subcommittees of the board—again, one subcommittee for each prize category. The subcommittees then recommend winners to the full board, which passes the recommendations on to the president of Columbia, who announces the winners. The board rarely rejects a recommendation from one of its subcommittees; the subcommittees rarely overturn the findings of the prize juries. Since 1975, when the trustees of Columbia University withdrew from the Pulitzer process, the president of the university has accepted all of the board's recommendations.

The Pulitzer Prize competition is open to anyone willing to fill out an entry form and remit the $20 entry fee. In journalism, the vast majority of entries are submitted by newspapers, which relish the prestige that comes with winning a Pulitzer. There is no limit to the number of entries a single newspaper can submit; in fact, most large dailies make multiple submissions: in 1990 New York *Newsday* led all comers with forty-one entries, followed by *The Washington Post* and the Associated Press, each with forty. *The New York Times* had thirty-five. A total of 1,770 entries were made that year. The same rules apply to the book awards. Not surprisingly, entries are dominated by large publishing houses. In 1990, 590 different books were submitted: 181 in general nonfiction, 123 in poetry, 115 in fiction, 92 in biography, and 79 in history.

The president of Columbia announces Pulitzer Prize winners at a press conference in early April; the actual judging begins long before that. For books, the process begins on December 31, the entry deadline. A copy of each entered book is sent by the Pulitzer Prize administrator, a Columbia University official, to each member of the five different book juries. Each book category has its own three-member jury, composed of college professors and writers with expertise in the category they oversee. History professors, for example, serve on the history book jury, and novelists serve on the fiction jury.

Jurists for each category have until January to compare notes, usually over the phone, and come up with three finalists to submit to the Pulitzer board. That means a total of fifteen books, three in each of the five book categories, is nominated for judging by the Pulitzer board. Rather than having every board member read all fifteen finalists between January and April, the board breaks down into seven subcommittees of three, with some members serving on more than one panel. There is one subcommittee for each of the five book categories.

Members of the subcommittees have until April to read the books and formulate their opinions. That way, by the time the whole board convenes at Columbia for two days in early April to judge the finalists, the subcommittees will be ready to recommend winners in their respective bailiwicks.

The drama and music juries, each composed of three individuals with expertise in the field they are judging—theater critics in the drama category, composers and music critics in the music category—submit three finalists apiece to the board on March 1. Members of the board's drama subcommittee then have a month to see the plays performed, or, if that is not possible, to read them; the music subcommittee listens to tape recordings of the three nominated compositions to reach its decision. Final recommendations are made to the board as a whole when it assembles in April.

The journalism awards work slightly differently. Each of the thirteen journalism juries has five members rather than three. The thirteen juries (one jury covers both photography categories) gather at Columbia for two days in early March to sift through all of the 1,770 or so entries in fourteen categories and come up with three finalists in each category. The sixty-five jurists, primarily working journalists, editors, publishers, and former Pulitzer winners, are not paid anything for their labor. (Book jurists receive a $1,000 stipend.) At the end of the two-day judgment period each jury nominates three finalists, which will then be submitted to the Pulitzer board for final judgment. Juries list their nominations in alphabetical order and do not indicate any preference among the three finalists. A total of forty-two nominations in fourteen journalism categories are made to the Pulitzer board.

When deciding among the journalism finalists, the board does not divide into subcommittees. Each final entry is read by all eighteen board members during the two-day judgment period in April. Winners are picked by a vote of the board. Since the board is composed of editors and publishers, many of whose newspapers are involved in the competition, rules have been adopted to prevent a conflict of interest. A board member must leave the room when entries from his newspaper are being discussed. A board member is not allowed to vote in any category that includes an entry from his newspaper.

The Pulitzer Prizes have been called the Academy Awards of journalism, and like the Academy Awards they have suffered their share of controversy. Since 1917, when the prizes were first given from a trust established by newspaper magnate Joseph Pulitzer in his will, many dubious choices have been made. In 1941 for example, Hemingway's *For Whom the Bell Tolls* was rejected on the grounds that it was "offensive and lascivious." The board awarded no fiction prize that year. More recently, in 1981, Janet Cooke of *The Washington Post* received the prize for feature writing for "Jimmy's World," a story about an eight-year-old heroin addict living in Washington, D.C., which included eyewitness accounts of "Jimmy" shooting up. It turned out that the reporter had completely made up the story, and the *Post* had to give up the award.

Despite the Pulitzers' long history of controversy, only two winners have ever refused their prizes: Sinclair Lewis, who won for his novel *Arrowsmith*, in 1926, and William Saroyan, who won for his play *The Time of Your Life*, in 1940. "All prizes, like all titles, are dangerous," Lewis wrote in rejecting his Pulitzer. Even so, he had no problem accepting a Nobel four years later.

How do they make crack?

TO MAKE CRACK, which is cocaine in a purified, hardened form, it is first necessary to make cocaine. The process begins in the Andes of South America, with the native coca shrub. (Incidentally, coca is a different plant entirely from the cacao tree, from which we get cocoa.) Coca leaves contain minute amounts of the alkaloid cocaine. To extract it from the plant, drug entrepreneurs soak the leaves in an acid solution composed of about a pound of sulfuric acid for every forty-five gallons of water. Cocaine workers accelerate the dissolution process by climbing into the vats and stomping the mash. Even so, the leaves must soak for twelve to eighteen hours. If all goes well, a brownish tea, called *caldo* (Spanish for "broth"), is produced.

The *caldo* is strained into a second vat, where the acid is neutralized by the addition of lime; then kerosene, an organic solvent, is stirred in. The mixture is decanted into a second acid solution, containing one and a half tablespoons of sulfuric acid per liter of water. Into the mix goes sodium bicarbonate, and voilà, a batch of coca paste is born.

To remove impurities from the crumbly paste, cocaine manufacturers dissolve it in acetone, heat it, and run it through a press. Once rinsed, coca paste is again diluted in acetone and then mixed with a solution of acetone, ether, and hydrochloric acid. The cocaine alkaloid reacts with the hydrochloric acid to produce a damp, powdery substance that remains at the bottom of the bowl after the ether and acetone are poured off. This is cocaine hydrochloride, ready for export as soon as it dries.

Before selling cocaine hydrochloride powder, which is 90 to 100 percent pure, drug dealers stretch the supply and thereby raise profits by cutting it with all manner of white substances, including talc, amphetamines, and cornstarch. That at least is the traditional method. In the mid-eighties, though, some enterprising drug lord hit upon a new way. He dissolved cocaine hydrochloride in a mixture of water and ammonia and then cooked it, to form crystalline rocks—potent, highly addictive crack. This drug can be smoked and thereby absorbed very quickly into the bloodstream, without recourse to the dangerous ether-fueled hookahs used to smoke, or free-base, regular cocaine.

In addition to being more potent than cocaine powder, crack is cheaper. On the street, a gram of cocaine costs between twenty-five and fifty dollars, but since a relatively small amount of cocaine hydrochloride can be converted into a sizable chunk of crack, prices are kept down. An individual hit of crack can sell for as little as five dollars, making it an affordable alternative.

?

How do they measure the unemployment rate?

EVERY MONTH, the Census Bureau surveys 59,500 American households, asking members about their current work habits. Once the data is in, the census takers ship it over to the Bureau of Labor Statistics, where analysts weigh the figures to project the national rate of unemployment.

To arrive at an unemployment figure, labor statisticians first determine the size of the work force, which equals all employed persons plus all unemployed persons. Children under sixteen years of age, prison inmates, and patients in mental institutions, sanitariums, and homes for the aged, infirm, and needy are excluded from the labor force. Not every jobless person qualifies as unemployed. Some are independently wealthy, some are retired, and others are just layabouts. Technically, by government definition, an unemployed person is "a civilian who had no work during the survey week, was available for work, and had made specific efforts to find employment during the prior four weeks." Laid-off workers who are waiting to be recalled to their jobs are automatically counted as unemployed, whether or not they are looking for alternative jobs. The unemployment rate is calculated by dividing the number of unemployed people by the number of people in the labor force. So, if survey figures show that the 59,500 sample households contain 80,000 workers and an additional 20,000 people who are of age and would like to work but have been thwarted in their attempts to find jobs, it means the work force stands at 100,000. By dividing the 20,000 would-be workers by the total labor force of 100,000, one arrives at the jobless rate, in this case 20 percent.

Strictly speaking, the jobless rate applies only to the 59,500 households that answered the government survey, but since the control group represents an accurate demographic cross section of the American public, the Bureau of Labor Statistics weighs the returns to arrive at a national number. Households from all fifty states and the District of Columbia are represented in the monthly sample, which covers 729 areas in more than 1,000 cities and counties. Any occupied living unit, excluding jails and long-term health-care facilities, is eligible for interview. A typical sample group might include single-family dwellings, apartment complexes, public housing units, and homeless shelters. People without any residential address, such as the unsheltered homeless, are not represented in the survey.

In order to keep the survey fresh, the sample households are constantly rotated. The total sample group of 59,500 households is broken down into eight subgroups, each of which is interviewed for a total of eight months. In any one month of the survey, one eighth of the respondents are new, one eighth are in their second month, one eighth are in their third month, and so on up to the final eighth who are in their final month of enumeration. The rotation system is designed to reduce statistical discontinuity by ensuring a 75 percent overlap of subgroups from month to month and a 50 percent overlap from year to year.

The survey is now a little over fifty years old. Begun in 1940, the business of counting the country's unemployed was originally a WPA project, conceived as a way to put people to work.

?

How do they interview and hire simultaneous interpreters at the United Nations?

THE UNITED NATIONS EMPLOYS 142 simultaneous interpreters, whose job it is to orally translate speeches at the very moment they are being delivered. A simultaneous interpreter listens to words spoken in one language and instantly relates in another language what is being said. Unlike translating a written document from one language to another, simultaneous interpretation is live—it is done almost instantaneously, without recourse to reference books or other aids. The work has been likened to "driving a car that has a steering wheel but no brakes and no reverse."

All UN speeches are simultaneously interpreted into each of the organization's six working languages: Arabic, Chinese, English, French, Russian, and Spanish. A UN diplomat who wants to use a language other than one of these six—Tamil, say, or Quechua—must provide his own interpreter.

The first step toward becoming a UN staff interpreter is to

take a comprehensive language test, which is open to anyone with a college degree in languages. The exam consists of oral and written sections that are designed to test speed, accuracy, clarity of accent, and familiarity with idiom. The UN administers tests when it decides it needs more interpreters. It may be given at any of the UN's sixty-seven information offices scattered around the world.

A job candidate who passes the test is granted a personal interview so that the UN may learn more about the candidate's character and background. Interpreters must be able to interpret all speakers accurately and dispassionately, regardless of their own political opinions. Since interpretation is more a scholarly exercise than a political one, candidates are not screened specifically for political biases. Such policing is not necessary: a staff interpreter who editorializes on the job will not last long at the UN. Although the interview is not, theoretically at least, a further test of language skills, the interviewer might well use more than one language during the session; once a person has passed the test, the superiority of his language skills is accepted without question.

A few things can happen to a candidate who successfully negotiates the interview. If there are staff openings, the candidate can be recruited for a job, sometimes just on a temporary basis to help out when the work load is especially heavy. If there are no immediate openings, he can be placed on a roster of qualified candidates from which future openings will be filled. UN officials have found, though, that many test takers are angling for an altogether different option: a letter of recommendation that says they have been positively screened for UN interpretation. That letter can be worth quite a bit more in the private sector than at the UN, where salaries for simultaneous interpreters start at just over $20,000 a year—scarcely a livable income in New York City.

?

How do they spread a computer virus?

A COMPUTER VIRUS IS a piece of written code that copies itself, spreads itself around, and tells a computer what to do. It might, for example, infect all the IBM-compatible computers it can, and on the next Friday the 13th, find all the decimal points in the data where it is and move them one digit to the right. Or it might simply destroy all the data and write "Gotcha!" on the screen.

Human viruses are spread by all sorts of contact. Anyone who works in an open office knows how fast a cold can spread, with everyone sneezing on each other, sharing telephones, shaking hands. Computers obviously don't sneeze. But they have plenty of chances to come in contact with each other, and every contact is a chance to transmit a computer virus.

Virus codes can be added to the lines of written code that make up computer programs, hidden among thousands of lines of commands that tell the computer what to do, and when. A computer hacker—one of those highly capable computer programmers who works day and night at his keyboard—could write a virus into a "bootleg" program before giving a copy of the disk to a colleague. Once the disk is inserted into a healthy computer, the virus copies itself onto other program files on the main computer and even into the computer's operating system, its brain.

The next time anyone inserts a disk into the infected computer to play a computer game, the new disk will be infected too. Or if the computer sends some program by modem, the virus goes along. And when the computer taps into a network, the virus sends itself, too. Traveling by network, a virus can infect hundreds of thousands of computers in just a few hours. Most viruses are started by one person who's up to no good, but very few are *spread* intentionally.

In 1988, a Cornell University computer science graduate student created a virus and released it into two networks that scientists use for exchanging information. The student meant only to show off his skills by gaining access to hundreds of research facilities' computers. But the virus went wild, replicating itself faster than the student ever intended and crippling computers all over the United States. The Computer Virus Industry Association estimated that $98 million was spent to clean up the damages left in its path.

Expert hackers have proven themselves capable of finding access to corporate and federal government computers and leaving virus code behind. Sometimes this is done through "trapdoors," the secret ways into a system that avoid the usual security measures like passwords.

In 1988 the Chaos Computer Club of West Germany gained access to NASA's computer and maintained access for five months before being discovered. Its members claim to have left behind viruses that will activate in the future, although NASA's own computer experts deny this.

Virus guards—codes written by the good guys—have helped curb the spread of viruses. They can detect typical virus language (and some can detect particular, known viruses) before the infected program runs on a computer.

?

How do they get the caffeine out of coffee?

THE BITTER, WHITE SUBSTANCE caffeine makes up only about one percent of the composition of coffee beans. The U.S. Food and Drug Administration requires 97 percent of that tiny amount to be removed before the label may say "decaffeinated," and some manufacturers do even better than that.

The difficult part of the decaffeination process, as any coffee lover can attest, is to maintain the rich flavor and aroma of regular coffee. Coffee beans contain more than a hundred chemical compounds, and it is virtually impossible to replicate coffee's aroma in the lab. During decaffeination, many of those compounds leak out of the bean. Getting only the caffeine out, and leaving those compounds in is the trick. And it's not just the consumer who benefits from good extraction processes: pure caffeine drawn from the beans is sold to pharmaceutical and soda companies.

The method of caffeine extraction most commonly used by manufacturers removes the caffeine by means of an "agent," or solvent. The agent used is either methylene chloride, which is also used in degreasing fluids, and ethyl acetate, a substance found in fruits.

The agent method can employ either a direct or an indirect process. In direct processing, green coffee beans are steamed for at least a half hour to soften them, open their pores and loosen the caffeine molecules. Then they are rinsed in the agent over and over again for hours. The solution and the caffeine are drained off and the beans are steamed again for several hours to drive out all traces of the agent, and finally dried.

In the indirect process, the green coffee beans are soaked in hot water for several hours to extract the caffeine; the water draws off a slew of flavorings, too. The watery solution, then, rather than the beans, is treated with an agent to remove the caffeine, and heated so the agent and the caffeine evaporate together. The rest of the solution and flavorings are returned to the beans and the beans are dried.

It's generally believed that almost none of the agent is left in coffee beans after either process is completed. Furthermore, roasting evaporates most of the few traces that are left. Still, consumer concern over the use of these chemicals has led some manufacturers to favor yet another method of decaffeination: the Swiss Water Process. This method calls for soaking the beans in water for a day, then running the watery solution through charcoal or carbon filters that sponge off the caffeine. Afterwards, the water is recirculated with the beans and the beans are dried. This takes longer than other methods, and some people think it doesn't return as much of the flavor to the beans as other methods. Other people say it has improved over the years. Anyway, a lot of consumers are willing to pay the extra cents per pound for what seems a natural way to decaffeinate.

Yet another method uses oils from spent coffee grounds. The beans sit for several hours at high temperatures in a container filled with decaffeinated coffee oils. Triglycerides in the oils pick off the caffeine in the fresh beans, which are then drained and dried.

Coffee companies are trying out a new method, using "supercritical" carbon dioxide, a pressurized liquid-gas similar to the carbon dioxide that effervesces in your soda can. Pumped through a vat of beans, the stuff seems remarkably capable of picking off the caffeine and then vanishing completely.

?

How do they test golf balls before offering them for sale?

THE UNITED STATES GOLF ASSOCIATION enforces strict standards governing the characteristics of golf balls. An official ball must not weigh more than 1.620 ounces; it must have a diameter of at least 1.680 inches; it must be spherically symmetrical; the velocity of the ball must not exceed 250 feet per second when launched by a USGA-approved mechanical striker; the ball must not travel farther than 280 yards when struck by a standard testing apparatus at the USGA proving grounds.

When golf ball manufacturers want to introduce a new product, they set up a pilot run to produce a limited number of new balls. The balls are tested at the manufacturer's in-house research and development labs. The tests determine what production changes must be made for the balls to meet USGA specifications. After passing in-house testing, sample balls are sent to the USGA headquarters in Far Hills, New Jersey, for official USGA testing. If the balls make the grade, the manufacturer calibrates the factory machinery to duplicate the approved ball and proceeds with a full production run, making the balls that will finally be stamped, packaged, and sold. Select balls from this run are tested for deviation; if any flaws are detected, the machines are readjusted until the production-run balls meet USGA requirements.

The USGA testing facilities include three indoor laboratories. In the first lab, balls are tested for spherical symmetry and for behavior when struck. The USGA impact standard holds that a ball should remain in contact with the driver clubface for

0.000450 seconds. To test this, an air gun fires balls at a metal plate and a computer records the amount of time the ball spends in contact with the plate and the velocity at which it rebounds. Conformity to the symmetry standard, which maintains that a ball must fly the same height, the same distance, and remain in flight for the same length of time regardless of how it is placed on the tee, is tested with the help of Iron Byron, a robotic golfer. Iron Byron has the ultimate golf body: one club connected directly to a mechanical shoulder, with no knees, hips, back, head, or other mobile parts to interfere with his perfect swing. He always hits the standard ball 280 yards, with a 6 percent margin of error. Since Iron Byron is perfectly consistent, any deviation in the flight of the driven ball must stem from the ball.

The aerodynamics lab consists of a wind tunnel, slightly longer than six feet, where balls are tested for flight properties such as backspin and wind resistance. The flight trajectory of a golf ball is the product of its dimples, which create an air flow pattern that reduces wind resistance, and of the backspin that is the normal result of hitting a ball with a club. To measure "lift force," the air flow pattern created by a dimpled, back-spinning ball, testers mount a golf ball in a vise-like device in the wind tunnel and buffet it with high-speed air streams that simulate the wind currents experienced by a golf ball in flight. By recording the effect of wind on the mounted ball, technicians are able to measure lift force and predict how it would perform under ordinary conditions.

The third lab, where size testing is done, is kept at a constant temperature of 23 degrees Celsius to prevent the balls from shrinking or swelling because of temperature changes. The size test consists of dropping balls through a ring gauge equal in diameter to the legal ball minimum—1.680 inches. A test victory is declared if fewer than twenty-five out of one hundred randomly dropped balls pass through the ring. This lab is also where balls are measured for initial velocity, or the speed of the ball off the clubface. To test for initial velocity, balls are launched into the air by a mechanical striker that has been adjusted to send the balls flying at a speed of 250 feet per second, the legal limit. The flight of the ball is measured photoelectrically, and any ball that exceeds 250 feet per second, plus or minus 2 percent, is deemed too springy.

After passing the indoor lab tests, the new batch of balls is taken out to the test range to be checked for conformity to the overall distance standard. Iron Byron strokes bucket after bucket of balls 280 yards straight down the fairway. If too many of the test balls resist Byron, either by soaring too far or falling short, the batch is discarded as faulty.

Golf ball manufacturers produce thousands of test balls, none of which are ever labeled or sold, before making a single ball for retail. The balls carried by sporting goods stores have not themselves been tested, but represent instead exact duplications of successful test balls.

?

How do they see a black hole?

THEY DON'T, ACTUALLY. A black hole is created by the collapse of an aged and massive star, and its gravitational force is so strong that it draws in even the light around it that would make it visible. And even if black holes were visible, their small size would still make them hard to see from Earth; some are only two or three miles across.

However, the location of black holes can be deduced by their strong effect on what's around them. If an ordinary star is near a black hole, for example, hydrogen and helium gases from the star's outer layers will spiral into the hole. As they do, they become highly compressed and heat up, giving off high-frequency radiation including X rays. Astronomers record the otherwise inexplicable radiation between the star and its invisible companion with sensitive satellites outside the earth's atmosphere. As many as six X-ray sources that almost certainly involve black holes in our galaxy have been detected with satellites.

If a very massive black hole is at the center of a galaxy or a star cluster, astronomers will also note increases in the velocities and densities of stars as they approach the area where the hole probably is. This is because the black hole's force, although it will not always pull the stars directly into itself, affects the movement and even the weights of the bodies around it.

?

How do they get the music onto a CD?

IN MARCH 1983 a new technology appeared on the U.S. music scene that would make the traditional LP record a mere item of nostalgia, a curio to be picked up at yard sales by baby boomers. The digital compact disc (CD) marked a radical departure in quality and convenience from the twelve-inch long-playing record.

The LP is an analog recording, which means it is an analogy or model of the actual music. Sound is stored in continuous grooves, whose side-to-side squiggles and variations in depth correspond to the original audio signal. A stylus running along the groove converts the mechanical motion into an electrical signal,

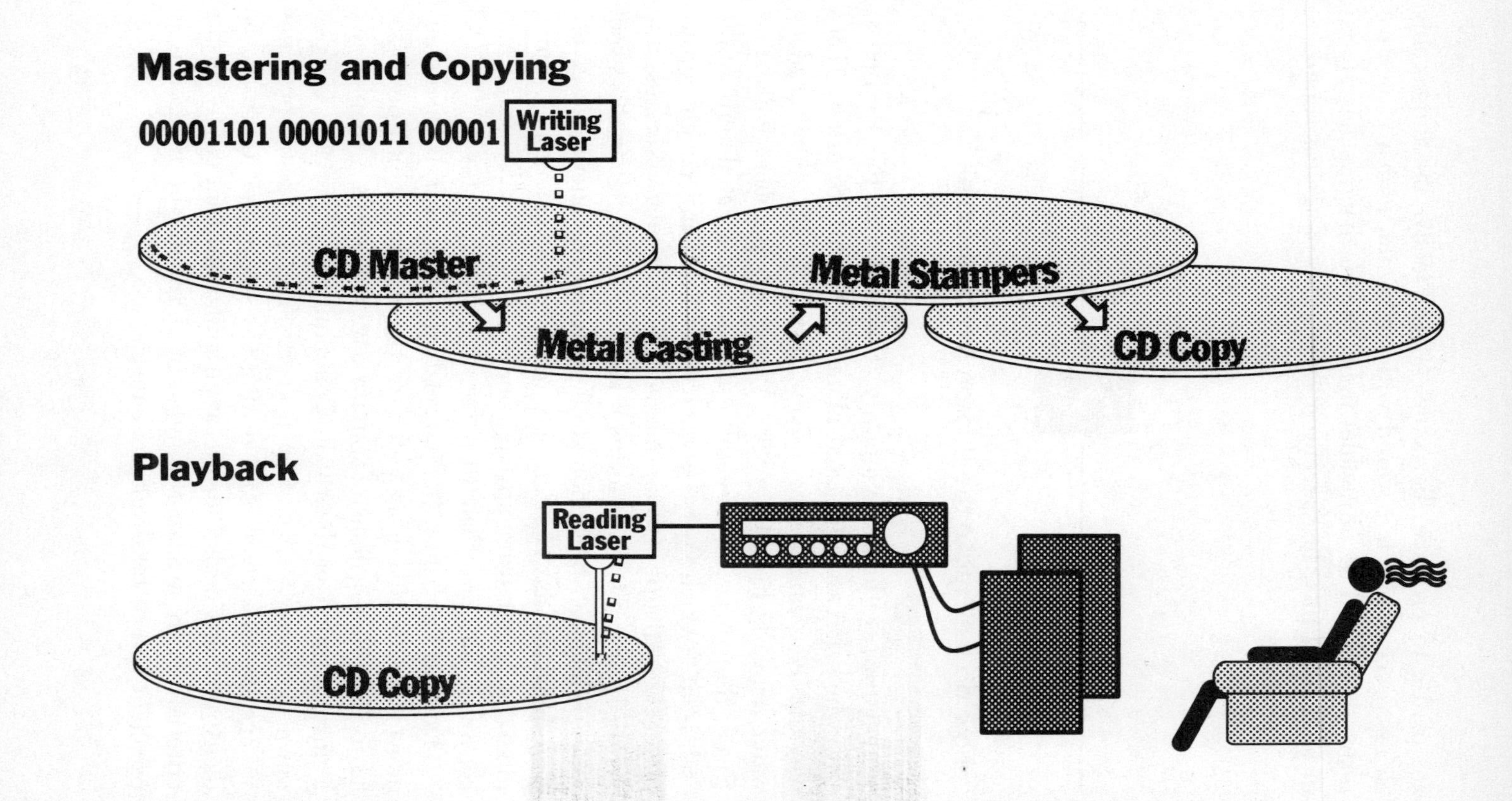
Mastering and Copying
00001101 00001011 00001
Writing Laser
CD Master
Metal Casting
Metal Stampers
CD Copy
Playback
Reading Laser
CD Copy

ate plastic to produce the actual CDs. Finally, the impeccably clean, precise disc is covered with aluminum and a protective resin, then dried by ultraviolet light. In your hand, safely stored in millions of pits, lies seventy-five minutes of ageless sound.

?

How do they make the Hudson River drinkable?

SURPRISINGLY, the first impurity to consider in the Hudson is salt, which is easier to avoid altogether than to remove. That's why the five communities in New York State that draw drinking water regularly from the river—Waterford, Highland Falls, Port Ewen, Rhinebeck, and Poughkeepsie—are all located north of the "salt front," which floats between West Point and Poughkeepsie. Here the Atlantic Ocean meets the Adirondack springs; the dense, heavy salt water forms a kind of wedge below the sweet water, which can be siphoned off for water treatment. New York City draws on the Hudson only in a drought emergency, but it pipes the water from upstate, and even then, river water may constitute no more than 10 percent of the supply.

Douglas Fairbanks, Jr., (no relation to the actor) is chief operator of the Poughkeepsie water treatment plant, which processes 11 to 12 million gallons of water a day. He claims, "People have a misconception of the river. Communities have really buckled down in the last ten or fifteen years and stopped dumping raw sewage into the river. And testing has shown that there are no dangerous chemicals there in any significant quantity. By all the criteria that are available to judge these things, the Hudson is an excellent source [of drinking water]."

But not everybody agrees with Fairbanks; the question lies in how you measure the water's purity. In the nineteenth century, civil engineers concerned themselves with bacteria and the spread of infectious diseases such as cholera and typhoid. Now scientists are studying the long-range effects of newer kinds of waste.

Says Cara Lee, environmental director of the civic organization Scenic Hudson, "The federal government doesn't have water quality standards for a lot of carcinogenic substances. Every year the Hudson receives 200,000 pounds of industrial toxins that are known or suspected to cause cancer in people: lead, cadmium, mercury, PCBs, pesticides, dioxins. The water-treatment plants test for bacteria and pathogenic organisms almost every day; they test for these toxins very infrequently, if at all." Lee cites studies showing that traces of these chemicals may cause cancer and disorders of the nervous system, and affect fetal development over the long term.

Susan Shaw, an environmental engineer with the Public Water Supply Section of the Environmental Protection Agency (EPA) in New York, doesn't deny Lee's claims. "The EPA can't regulate every substance in the river. We have standards for the most common ones," she explains. "There are always going to be some pollutants where there are industries discharging."

The same spirit of compromise underlies the EPA's stand on chlorination. When chlorine atoms bond with naturally occurring acids in the water, they form trihalomethanes (THMs) such as chloroform and bromoform, which are associated with cancer of the intestine, bladder and rectum. "We'd like to see a balance," states Shaw. "You need enough chlorine to provide adequate disinfection for waterborne diseases, which are an acute risk to health. A whole town could get gastroenteritis. But we have to weigh the long-term chronic exposure to disinfection by-products."

Apparently the EPA has been experimenting with filtering out natural acids before chlorination and with substituting ozone as a disinfection agent, but the agency aims to lower the levels of

THMs, not to eliminate them. Right now the EPA allows up to one hundred micrograms of THMs per liter; some activists are trying to reduce that figure to twenty. How many micrograms per liter in the Poughkeepsie plant? Fairbanks says sixty to sixty-five.

Here's how Poughkeepsie's traditional water filter system works:

The river water is pumped into floculators—enormous, rectangular concrete basins divided in two, each with a million-gallon capacity. The plant supervisor adds a mixture of chemicals that includes alum. The chemicals attract the sediment and particles in the water, which include such pollutants as bacteria, viruses, rocks, plant matter, and sewage. The pollutants and chemicals clump together and settle at the bottom of the floculators. Two to four hours later, the water is pumped into the second half of the basin as a pipe feeds in a precise flow of chlorine to kill any remaining microorganisms.

The water from the basins flows through a filter to remove the tiniest particles which have escaped coagulation. The filter is composed of blocks of perforated tile that hold two or three graduated layers of stone; the stone ranges in size from peas to golf balls, getting coarser toward the bottom. Next, the plant supervisors add fluoride and also treat the water for "corrosivity," by mixing in lime, a process that makes the water more alkaline so that it won't eat into pipes and release toxic metals like lead.

And then you have drinkable water. Or do you?

?

How do they pack Neapolitan ice cream?

DESPITE WHAT CERTAIN NOSTALGIA-HAPPY entrepreneurs would have you believe, ice cream production is a big, industrial business. The United States produces more than 900 million gallons of ice cream every year. During the first half of this century ice cream tycoons opened vast plants where noisy stainless steel machines replaced the old hand-cranked wooden ice cream freezer. In addition to increasing productivity, the new equipment allowed for the perfection of the Neapolitan brick's tri-flavored parallelism.

The three individual elements that together make a carton of Neapolitan—chocolate, vanilla, and strawberry—are frozen in separate ice cream freezers. When the ice cream attains a stiffness desirable for packing, the three freezers pump their flavors into distinct compartments of a rectangular device called a header. The header molds the ice cream into a manageable brick, and then the contents of all three compartments are simultaneously extruded into a waiting box. Because the flavors settle differently in the box, surface area is not divided equally among the three flavors.

The label Neapolitan was originally used to designate a layered ice cream that included Italian ice, preferably orange or lemon, among its constituent parts. The people of Naples, it seems, were pioneers in the field of ice cream development. Marco Polo brought the original recipes for water ices to Italy from China in the thirteenth century. Neapolitan, by the way, is now the fourth most popular ice cream category, surpassed only by vanilla, candy confection mixes, and chocolate.

?

How do they make sure condoms won't break?

A ROMANTIC INTERLUDE isn't the only time people worry about condoms breaking. Manufacturers worry about it all the time. So they put the condoms through elaborate tests to make sure they're leakproof and strong.

The key ingredient in the success of the multimillion-dollar modern condom industry—Americans buy some 420 million of them a year—is latex. It's strong, light, and malleable. The manufacture of condoms begins with huge vats of liquid latex, containing anywhere from 250 to 1,000 gallons. The liquid is tested for consistency and filtered through cheesecloth to remove impurities. Once a batch of quick-drying liquid latex is ready, machines dip glass condom molds, called mandrels, into the opaque, viscous fluid. The liquid latex adheres in a thin film to the cylindrical mandrels; the film dries when exposed to air. After the film of latex has dried, the mandrels are dipped into the vat a second time to achieve a double thickness. The process is something like taking a plaster of Paris mold of an object, except in this case the molding material is thin and elastic.

After condoms have been shaped they're stretched by 15 percent onto large steel molds, and tension and strength tests are performed. The condom-encased steel molds are dipped into a vat of charged water. If the electric charge passes through the condom to the steel rod, the condom is discarded as faulty. In a second test, pumps fill the condoms with water to make sure they don't leak. A number of condoms from each production batch are also randomly selected to undergo the inflation test: condoms are blown up like balloons to see how much pressure they can withstand. Industry guidelines specify that condoms have a breaking volume of an almost unbelievable twenty-five liters or more.

Condoms that pass the battery of tests are rolled off the mandrels and hermetically sealed in foil. Depending on the model, a water-based silicon lubrication might be applied to the product before it is packaged. The tensile strength of latex allows condoms to be rolled off the smooth rods as easily as a sock can be rolled down a leg. On the fully automated production line the rolling is done by cylindrical devices that slide down the mandrels, pushing and rolling the condoms free. Human workers monitor the system for glitches.

Latex, which is made by mixing synthetic chemical compounds with the sap of the rubber tree, was developed in the 1930s. Coupled with mechanized production, the new material made inexpensive, reliable condoms a practical reality for the first time ever. In the late nineteenth and early twentieth centuries, condoms were commonly made of vulcanized rubber. The old material was relatively bulky and, as they say in the business, unresponsive.

Vulcanized "rubbers" were, however, less expensive than natural-membrane condoms, which have been around since at least 1350 BC, when Egyptian men of the upper class wore them as decorative sheaths. A natural-membrane condom is made from a sheep's cecum, an intestinal appendage similar to the appendix. You can make only one condom per sheep, and the production process is laborious, both factors contributing to the expense of the item. Such devices were commonly used by men of means right through the eighteenth century. In fact, natural-membrane condoms are still available, at a cost of about four times that of latex models. Advocates claim they are more sensitive and conduct heat better.

Recent condom innovations include adhesive strips around

the open end and large-size condoms. The adhesive condom was developed by Carter-Wallace, whose Trojan brand is the U.S. best-seller. Company officials say the adhesive band will prevent slippage, thereby providing additional comfort and security. A number of companies are currently manufacturing bigger-than-average-sized condoms, the norm being seven and a half inches long and 2.08 inches in diameter. Although there apparently isn't any great physiological need for them, research done by the Kinsey Institute indicates some psychological need might exist: 70 percent of the men contacted in one poll said they would be happy if their condoms were larger.

?

How do they make stainproof carpets?

E. I. DU PONT DE NEMOURS AND COMPANY, headquartered in Wilmington, Delaware, was among the first companies to come up with a new technology for making stain-repellent carpets. In the early eighties one of du Pont's research chemists, Armand Zinnato, was developing a nylon fiber that could be dyed at room temperature. He worried that such a fiber would be very susceptible to stains and started applying to the nylon certain agents formerly used to help keep a product colorfast when washed. These agents proved effective in resisting all manner of stains from food.

In September 1986, du Pont launched Stainmaster carpets, guaranteed to resist food stains and even accidents by the dog. How does it work? Before the carpet is made, the nylon fiber is dyed. During the dying process, technicians heat the fiber to a high temperature and then apply a top-secret stain-resistant chemical, which seals the microscopic pores of the fiber. Just as a sealant on a molar prevents tooth-harming bacteria from getting at the tooth, so the chemical envelops each fiber in a protective sheath. Spills merely sit on the nylon's surface and so can be mopped up with a sponge, just as the commercials demonstrate.

Later in the manufacturing process, after the carpet has been made, the nylon is sprayed with fluorocarbons that provide surface protection against soil. After much wear and tear, this treatment rubs off, but the stain-resisting chemicals locked into the fiber continue to work.

?

How do they teach a bear to ride a bicycle?

BEAR CUBS AND HUMAN CUBS are built in kind of the same way: they're sturdy, fairly well balanced, and capable of walking upright. In addition, they both put their heels down first when they walk. So for trainers at the Ringling Brothers and Barnum & Bailey Circus, where bears have been performing since the turn of the century, teaching a cub to ride a bike is almost like taking a kid out with his first two-wheeler.

The bear is given a small bicycle with training wheels and a custom seat to fit its bearish behind. Stirrups hold its rear paws on the pedals; the bear leans forward with its front paws on the handlebars. The trainer pushes or gently pulls the bike to familiarize the bear with the pedals turning. Once the bear can pedal pretty well itself, the training wheels are lifted. The bear's first solo run with no side wheels is bound to include a few spills, but it has thick fur and decent balance, so it gets few bumps and practically no scraped knees.

A bear cub can be frustrating during these training sessions because it's so playful at this age and its attention span is short; it doesn't always feel quite like staying on the seat. But the cub can be coaxed into compliance with food and praise. Trainers say that bicycling isn't as foreign to bears as people may think. If the cub trusts its trainer and if it gets lots of positive reinforcement (i.e., more food and praise), then its riding a bicycle is no different from a dog darting after a Frisbee. Bears also can learn to juggle, jump rope, bounce on a trampoline, walk a tightwire and ride a motorcycle.

?

How do they measure the ozone layer?

CONCERN ABOUT THE DEPLETION of atmospheric ozone really took off in the mid-eighties when scientists of the British Antarctic Survey first detected an ozone "hole" over their research station on Halley Bay. They found that the springtime quantity of ozone over Antarctica had decreased by more than 40 percent between 1977 and 1984. The "hole," or area of extreme ozone depletion, was even wider than Antarctica and, extending from about seven and a half miles above earth to a little over fifteen miles above, spanned most of the lower stratosphere where atmospheric ozone is concentrated.

Ozone is simply a form of oxygen with a slightly different chemical composition: it comprises three atoms (O_3) rather than two (O_2). Ironically, this substance is toxic to human beings and other animals when it occurs at the lowest levels of the atmosphere. In the stratosphere, however, ozone is essential to maintain life of any kind on Earth. The ozone layer not only protects human beings from skin cancer by screening out harmful ultraviolet radiation, but it also protects the most elementary organisms in the food chain (bacteria, algae, protozoa) from lethal damage to their genes. The largest concentration of ozone is found in the lower stratosphere, from about nine and a half to nineteen miles above earth (fifteen to thirty kilometers), where it attains more than a thousand times the normal peak concentration in the air we breathe. Chemicals that dissociate ozone to oxygen by knocking off its extra molecule—namely nitrogen oxides and chlorine, each atom of which has the ability to destroy multitudes of ozone molecules—can remain in the stratosphere for years, devastating the ozone layer.

How have scientists learned so much about the ozone layer? Basically, through instruments on the ground, in balloons, in airplanes, and on orbiting satellites.

Ground-based equipment measures solar radiation at slightly different wavelengths, some of which are known to be readily absorbed by ozone, and some of which are known to pass through it. If, over time, scientists find there are more of the wavelengths that can be absorbed by ozone than the kind that cannot, they deduce that ozone is decreasing. Conversely, if absorbable-wavelength radiation decreases, ozone has increased. From such measurements the thickness of the ozone layer is calculated.

In 1987 a map of the ozone cover and hole over the Antarctic was created on the basis of data collected on board NASA's *Nimbus 7* satellite by the Total Ozone Mapping Spectrometer (TOMS). At the same time, 150 scientists drawn from around the world participated in the Airborne Antarctic Ozone Experiment in which specially outfitted, medium-altitude, long-range DC-8 and high-altitude NASA-ER2 airplanes measured the ozone-depleted region's size and chemistry. In addition to the renowned "hole," they found more widespread ozone loss over the Southern Hemisphere and markedly higher amounts of chlorine monoxide (a constituent element of the chlorofluorocarbons—CFCs—that are believed to be major culprits in ozone destruction) than those at other latitudes. They also found severely depressed levels of nitrogen oxides that can protect ozone by hindering chlorine's ability to attack it. A polar-orbiting satellite continues to track the development of the ozone hole over time, relaying its information to the Goddard Space Flight Center in Greenbelt, Maryland.

Long-term records show that ozone levels over the Arctic ice-cap have dropped roughly 5 to 6 percent over the last seventeen years. Recently a hole has been observed over the most northern latitudes in the spring. A team of one hundred scientists, with over $10 million funding supplied by NASA and NOAA (National Oceanic and Atmospheric Administration), has just completed a study of this hole and the role of man-made CFCs in causing it. With the aid of airplane- and balloon-borne instruments, the research team found that patches of stratospheric air contained up to fifty times the normal amounts of CFCs' chemical constituents and a significant absence of the protective active nitrogen.

On the basis of the vast bodies of data that have been painstakingly accumulated by scientists, NASA experts now predict that the hole in the Antarctic ozone layer could repair itself by the year 2100. They stipulate, however, that for such repair to occur there would have to be a worldwide phaseout of practically all emissions of CFCs and halon gases by the year 2000. Although the European Community countries are fully behind this and the United States almost as much so, several large industrializing nations have yet to commit themselves.

?

How do they count calories in food?

ENERGY IS TRAPPED in the molecular bonds inside food, and used by the body for all sorts of physical activities: shopping, studying, jogging, reading this book. A calorie is a way to measure that energy—both as it is supplied, and as it is burned.

Energy is discussed in terms of heat. One calorie—the term *calorie* here actually referring to a *kilocalorie*—is the amount of

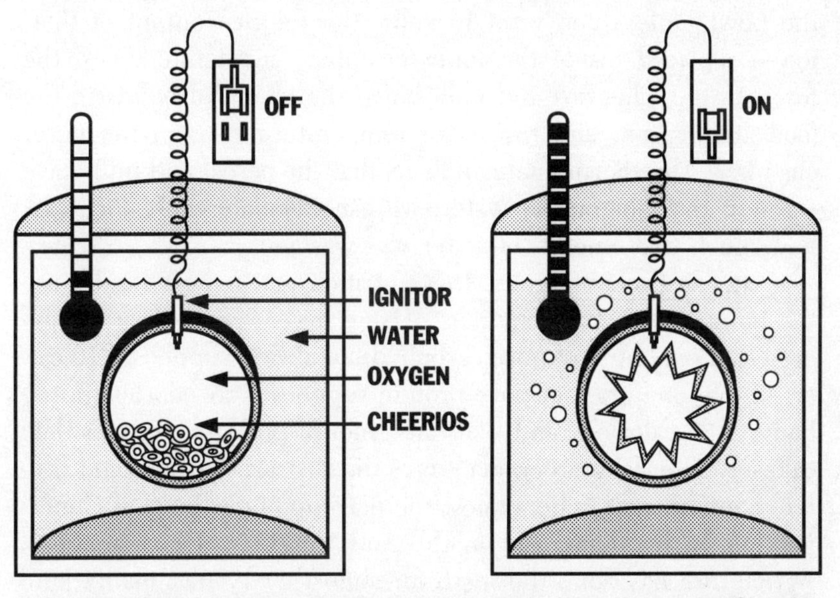

This bomb calorimeter measures the calorie count in food. Here, some Cheerios are placed in the inner chamber, or "bomb," which is filled with oxygen. The cereal is ignited and burns, increasing the temperature of the water in the outer chamber. The amount of temperature increase is used to calculate the calorie count, or the amount of energy contained in the food.

heat it takes to raise the temperature of one kilogram of water (about a quart) by one degree Celsius. When you read that a bowl of Cheerios and milk has 150 calories, it means that this breakfast contains the energy to raise about 150 quarts of water one degree Celsius.

Obviously, if you added a bowl of Cheerios to a kilogram of water it wouldn't automatically raise the temperature. You have to break the chemical bonds to first release and then to measure the energy. Your body's cells break the bonds and store the energy, and so does a piece of equipment called a bomb calorimeter.

A bomb calorimeter is small enough to sit on a table. It has an inner chamber called a "bomb," and an outer chamber filled with water, and both are surrounded by insulation. The bomb contains only oxygen and an electrical igniter. A thermometer measures the temperature of the water in the outer chamber.

An item of food—for example, that bowl of Cheerios (*without* the bowl, unless you want to know the calorie content of that, too)—is placed inside the inner chamber, and ignited with the fuse. As the Cheerios and milk burn, the chemical bonds in the food are broken, and the water temperature rises in the outer chamber. The thermometer tells us that the cereal and milk have warmed the kilogram of water; we can calculate the calories by how much that amount of water was warmed.

You can probably guess that nutrition researchers hardly use the bomb calorimeter anymore. Using it years ago, they discovered that one gram of pure carbohydrate always supplies 4.10 calories; that one gram of pure protein supplies 5.65 calories (in the body, 4.35 calories), and that one gram of pure fat supplies 9.45 calories. After laboratory analysis of the fat, carbohydrate and protein content, researchers know the makeup of our bowl of Cheerios simply by doing the math. And the U.S. Department of Agriculture has done the math for almost every imaginable kind of food already, and has published the results.

Incidentally, one way to measure the energy used by a person (rather than eaten by a person) is in a device similar to the bomb calorimeter, but bigger. A person working in the inner chamber burns off calories and the air gets warmer. (Have you ever been in an aerobics room after class? Same thing.) The energy raises the temperature of water in pipes located at the top of the chamber, and we can see how many calories have been burned.

?

How do they know how much money will come out of a cash machine?

WHEN YOU HUSTLE TO the bank for some quick cash, it's easy to wish that that anonymous machine would just once spit out a few extra bills, or accidentally hand out fifties instead of tens. But such fortuitous errors are rare.

Why? Inside an automatic teller machine, or ATM, are bricks of cash in various denominations, stacked in open-ended boxes with up to three thousand bills in each box. When you punch in your secret code and your withdrawal request—of, say, one hundred dollars—a computer in the machine sends signals to your bank's computer, telling it who you are and how much cash you want. If you're good for the money, the computers activate mechanical arms inside the teller machine, and they instruct the arms which bills to select and how many of each. In this case, the computer might designate five twenty-dollar bills.

At the end of the mechanical arm is a suction cup similar to that on a rubber dart toy. A vacuum hose within the cup strengthens its grip. Each arm picks a single bill off the top of the cash brick and passes it into a set of rollers. The arms' movement and suction make the noises you hear while you wait. A less-common system contains rollers that use friction to peel bills off the cash brick. As the bills roll one by one through the conveyor system, they pass between a light beam and a sensor, which measures the typical opacity of a U.S. twenty-dollar bill. (Using the opacity test, ATMs sometimes throw dirty bills into the bin, so banks prefer to use new money if they can get it.) Some machines have a sensitive set of rollers that measures the thickness of single bills as they pass through it. Either way, the machine can calculate how many bills you are getting. If it's more than your allotted one hundred dollars, the conveyor dumps all the money into a bin and tries the transaction again.

If, on your turn, an ATM runs out of twenty-dollar bills, then it will probably tell you it is out of money altogether. However if the machine has other denominations—ten-dollar bills, for example—then the next person to request a withdrawal will be given his cash in tens. So if the teller claims to be dry at first, be sure to try your withdrawal one more time.

?

How does the Pennsylvania Department of Agriculture get mentioned on cereal boxes?

IF YOU'VE EVER FOUND yourself alone at the breakfast table without a newspaper, you've probably resorted to reading the cereal box for diversion. There, along the side of the box, buried deep within the fine print, lies the cryptic notation, "Reg. Penna. Dept. of Agri." What, you may have wondered, does the Pennsylvania Department of Agriculture have to do with your box of cornflakes, made in Michigan and bought at your corner store?

The answer dates back to 1933, when the state of Pennsylvania enacted the Pennsylvania Bakery Law. The law was a direct response to the technological revolution of the early twentieth century, which enabled large manufacturing concerns to flood local markets with their goods. Baking became a big business with resourceful state, national, and even international baking companies supplanting small neighborhood bakeries. Pennsylvania passed its law to ensure that baked goods arriving from out of state met the same high standards as those produced locally. It stipulated that in order to peddle baked goods—including everything from potato chips and pretzels to pasta—in the state of Pennsylvania, companies must hold a Pennsylvania bakery license. Then, as now, to obtain a license a bakery must pass an annual inspection for cleanliness and labeling accuracy, and its employees must undergo yearly medical examinations.

Rather than try to visit each out-of-state bakery personally, officials at the Pennsylvania Department of Agriculture rely on the agriculture departments of other states and countries to annually inspect bakeries under their jurisdiction and submit their findings to Pennsylvania. The notation on the box stands as proof that the product within has met required standards and is licensed to be sold in Pennsylvania. Most bakeries find it easier and less costly to include the notation on all packaging, rather than specially printing just those boxes destined to wind up on grocery store shelves in Pennsylvania.

Pennsylvania is not alone in requiring bakery inspection and licensing—in fact, all fifty states now do so. But only Pennsylvania requires an inspection notation. Since standards still vary from state to state, the mark continues to ensure the quality of baked goods sold in the Keystone State. Recently, however, there's been a move afoot at the U.S. Department of Agriculture to standardize bakery licensing. If approved, the new rules could spell the end for Pennsylvania's unique notation. Naturally, officials at the Pennsylvania Department of Agriculture would hate to see it go. It's good advertising, they say, for a state that since 1933 has been certified pure.

?

How do they detect wind shear?

UNLESS YOU SPEND A LOT of time flying airplanes, you probably haven't given a great deal of thought to the subject of wind shear. Wind shear is the phenomenon of radically differing wind velocity and direction at slightly different altitudes. The occurrence of an adjoining northerly twelve-knot wind, say, and a southerly twenty-knot wind would constitute wind shear. It is a phenomenon of grave concern to pilots. Wind shear causes the aircraft to feel turbulence and lose speed; if the differences between the adjoining winds are great enough, the effect can be extremely detrimental to sustained flight. Wind shear is especially dangerous when it occurs at a low altitude in the vicinity of an airport, where it can affect planes in the vulnerable takeoff and landing phases of flight. Wind shear can cause a low-flying, slow-moving plane to lose airspeed and lift, sometimes so suddenly that the pilot does not have enough time to adjust to the new conditions and the plane crashes. Although wind shear can occur at any altitude, a plane flying high and fast will usually have plenty of time to compensate for its negative effects. For this reason, low-level wind shear—wind shear below roughly 800 to 1,200 feet—is of greatest concern to flyers.

Many airports in the United States are equipped with the low-level wind-shear alerting system (LLWAS), which detects weather conditions likely to produce wind shear. The system consists of a network of anemometers—fancy weathervanes—arranged to monitor wind speed and direction at various points around the airport. Data received by each of five or six anemometers on the perimeter of the airport and one at its center are fed into a computer in the control tower. If the computer detects strong discrepancies, it sets off an alarm. Control tower officials are then able to send out a wind shear warning.

A variety of meteorological phenomena can cause wind shear. Any kind of frontal zone, the boundary between different air masses, can produce it. Fast-moving cold fronts create prime wind shear conditions as they overtake slower warm air masses.

Thunderstorms are another wind shear breeding ground. They contain strong downward currents of air, called downdrafts and microbursts, that strike the earth and fan out horizontally. An airplane flying into a downdraft produced by a thunderstorm will experience all sorts of unenviable events. First, as the craft enters the downdraft, it meets with headwinds created by the flow of air across the earth's surface, and its lift and airspeed are increased. Then, as the airplane continues across the downdraft, downward currents become more pronounced and the plane is pushed toward the ground—it loses lift. At the same time, a rapid change from headwind to tailwind conditions causes the plane to lose airspeed, a situation that further erodes lift. And all of this can transpire in a matter of seconds.

Since wind shear often accompanies specific types of weather, experienced pilots can usually be alerted to it by checking for fronts and thunderstorms along the flight route, even before taking off.

?

How do they make fake photos on magazine covers?

IF YOU HAVE EVER READ *Spy* magazine you are probably familiar with doctored magazine cover photos. The satiric New York monthly once ran, for instance, a cover that included a picture of Henry Kissinger wantonly dancing the hula. It was not really a picture of Dr. Kissinger. Rather, it was a picture of Dr. Kissinger's laughing head attached to a picture of the jiggling body of an overweight dancer. The fusing of the two pictures was done so seamlessly, however, producing such a strong photographic likeness, that it is almost impossible to believe that Henry Kissinger did not actually do the dance we see him performing. *Spy*, of course, is not the only magazine that runs doctored photos on its cover; it just uses them to funnier effect than most.

Any picture—from a Monet painting to a Polaroid snapshot—is just a collection of different colors, organized in such a way as to create a likeness of a three-dimensional object. With the aid of new computer technology the colors of a picture can be rearranged in such a way that a new picture is created from the colors of the original. Fake magazine covers are made in photo studios by feeding an original image (a photo of Henry Kissinger, say) into a computer scanner, the desktop version of which is roughly the size and shape of a laser printer. The scanner records the photograph's picture information (colors and color placement), translating it into a digital code. The digitally coded picture, an electronic document, is transmitted to a computer, where it can be called up on the monitor as an image of the original photograph.

Think, for example, of a jigsaw puzzle. When assembled properly, the five hundred individual pieces create a large, cohesive picture. What a scanner does is turn an ordinary picture into a jigsaw puzzle. It isolates pixels, which are like electronic jigsaw puzzle pieces, except that each one is only the size of a pinhead. This jigsaw puzzle can be reassembled in any fashion—the pieces can be made to fit anywhere. Once the scanner has broken down the picture into its composite elements, the elements are manipulated into new shapes, making it possible to create a new, electronic image by rearranging elements of the original picture.

Unsightly wrinkles can be removed from a model's face by replacing the pixels that contain wrinkle information with smooth-skin pixels. Shadows can be deleted, backgrounds altered, celebrities decapitated, their disembodied heads reattached to other bodies. A tiff, that is, a hard copy of the new image as it appears on the computer monitor, can be printed by a color offset device. The finessed picture can then be reproduced on the cover of a magazine.

While a picture may be worth a thousand words, in the new age of computer technology there is no guarantee that those words are not lies. Any person with the right equipment can create a convincing, fake picture.

?

How do they make shredded wheat?

A STAPLE ON THE American breakfast table for decades, good old wholesome shredded wheat dates back a century. The story goes that at a small-town hotel in Nebraska in 1892, lawyer and inventor Henry D. Perky was suffering from indigestion one morning when he noticed a fellow guest eating an oddity for breakfast: boiled wheat covered with milk. The stranger, a fellow indigestion sufferer, claimed his bland dish cured their common ailment. This inspired Perky to set out to produce the stuff and to make it more palatable.

With the help of his brother in Denver, Perky devised a machine to extrude boiled wheat paste in stringy filaments, which were then woven into pillows of "shredded wheat." After leaving them for a day to settle, the inventor then baked them to keep them from spoiling.

The first commercial shredded wheat biscuits were produced by a Denver cracker bakery in 1893. Their intrepid inventor then headed east to promote their healthful virtues to the mass market. In 1901 Perky opened a conspicuous red-brick Shredded Wheat bakery at Niagara Falls, New York, dispensing free samples of the "original Niagara Falls cereal" to its thousands of annual visitors.

Whether because of its curative qualities or simply its novelty, Perky's shredded wheat caught on in a big way. In 1928 a large-scale manufacturer, the National Biscuit Company (predecessor of RJR Nabisco), bought the company, opening two new plants and introducing modern processing and packaging methods.

Today each plant processes daily several thousand bushels of soft white winter wheat that has been stored in adjacent grain elevators until needed. After the wheat has been cleaned and moistened in hot water, it is cooked in immense kettles for over half an hour. This 100 percent wheat gruel is then run through driers and cured in steel storage tanks to equalize its moisture content so as to yield a cool, dry paste of uniform consistency that can be shredded mechanically. The shredded filaments are then stacked up in layers on a moving strip which passes under a cutter that separates the straw-like material into biscuits. These glide on to the oven to be puffed up and baked crisp and golden before being sealed in moisture-resistant packages and shipped to grocery stores.

In addition to the traditional "big biscuit," shredded wheat is now made by competing manufacturers in snackable and child-friendly spoon sizes. And to please health nut and child alike, the wholesome wheat can be found bolstered with the bran of oat or wheat—or doused in sugar frosting.

?

How do they write words across the sky in airplane exhaust?

WHEN WE THINK OF skywriting, most of us picture a small airplane looping-the-loop to scrawl advertising slogans or "I love you's" with a vanishing vapor trail. But such laborious acrobatics are almost a thing of the past. Since the wind can wipe out a letter even before the daring aviator finishes it, an imaginative skywriter has dreamed up a longer-lasting, if less romantic, method of "skytyping." Although you occasionally see the original form of skywriting in this country, now you are more likely to see messages rapidly "typed" out across the sky in dot matrix form by a close formation of five or six airplanes.

The inventor of traditional skywriting is thought to have been a British flying ace, Major John C. Savage, who got his inspiration from the use of airplanes during World War I to spread smoke around ships and troops to camouflage them from the enemy. In 1922 he wrote "HELLO" to the crowds at a British racetrack. Later that year, a plane traced "HELLO, U.S.A." over the Polo Grounds in New York during a World Series game and followed up with: "CALL VANDERBILT 7200." Those who did reached the first commercial skywriting company in America, headquartered at the Vanderbilt Hotel.

Advertisers took to skywriting in a big way, but the business took a nosedive in the late 1940s, when television became a competitor. Despite the ingenuity of some pilots in holding their audience's attention—like the one who spent hours misspelling the name of a local furniture company in ever more creative ways—advertisers perceived TV as more efficient in getting across their message.

U.S. Navy pilot Andy Stinis, celebrated as the most skilled American skywriter, saved his vocation by patenting "Skytyping" in 1949, an innovative method of writing in the sky designed to make the medium faster and more versatile. Skytypers can write twenty letters in the time an old-fashioned skywriter can manage one. For instance, to write the letter *M*, a solo skywriter has to put his aircraft through these time-consuming maneuvers:

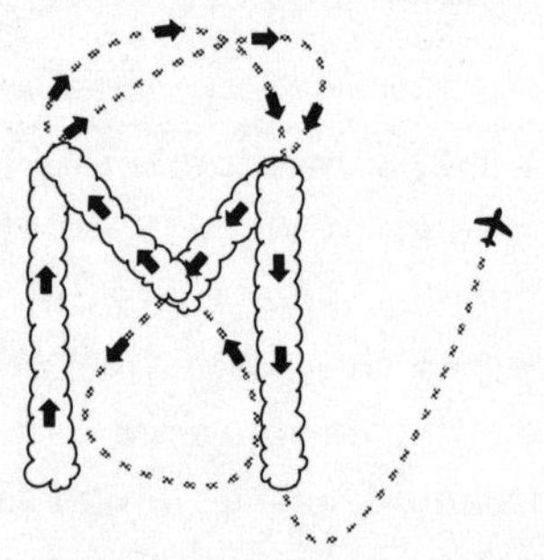

A "front-line" formation of five skytyping aircraft only needs to fly straight ahead, maintaining a distance of 225 to 250 feet between planes. When commanded by computer, they emit brief bursts of smoke according to this format:

Messages to be skytyped are transposed onto a tape as a sequence of punch marks along the lines of the diagram above, indicating when each aircraft should emit smoke. When the planes are at ten thousand feet and over their prospective audience, the tape is fed into a specially designed computer housed in the lead plane of the squadron. The computer reads the tape and triggers the lead plane's radio transmitter to send out radio signals on different frequencies specific to the radio receivers of each other plane in the formation. The radio signals activate each receiver to command a valve to open and start the process that releases a burst of smoke. The entire system is rather like an antiquated nineteenth-century piano roll.

To keep the typed-out message hanging in the sky for as long as possible, a patented solution is added to the planes' exhaust that makes it more visible and holds it in place longer. On a windless day, a skytyped message can be read in its entirety for as long as twenty-five minutes, as opposed to the few minutes of a conventionally skywritten one.

The exact recipe for Stinis's patented solution is a secret, but essentially it is a highly distilled form of paraffin gas, 100 percent biodegradable, nonpolluting and EPA-approved. Each time a plane's receiver is activated by radio signal, a valve releases a small amount of the distilled mineral oil into the plane's extremely hot (2,000 to 2,300 degrees Fahrenheit) manifold. The oil is instantly vaporized and forced out through a ten-foot pipe running along the top of the starboard wing and extending six inches beyond it. When the very hot vapor hits the very cold air at ten thousand feet, it condenses as a little, intensely white, artificial cloud, its steam molecules held together by the paraffin.

Skytyped messages usually run a thousand feet tall and six to eight miles long. With six aircraft the letters can be made even taller. The writing is visible for four hundred square miles; over a crowded urban area a message can be seen by as many as three and a half million people, and a single plane formation can print out thirty messages a day. Commercial skytyping has included pictures of diapers and diamonds strung out across the sky, and the aerial writers have taken on all sorts of personal missives, too. Sometimes skywriting the old-fashioned way with a solo plane (for no extra charge) they have proclaimed "ALICE DIVORCE ME.

GOOD-BYE HOUSE. HELLO ALIMONY" and, in another case, "REPENT. THE END IS NEAR." But the sky must be blue and the wind below thirty-five knots (forty miles per hour) or no one will see the message.

?

How do they determine who gets into Harvard?

MORE THAN 14,000 hopeful students apply for admission to Harvard each year, competing to fill out a freshman class of about 1,600. In an average year, roughly 15 percent of the applicants will be accepted; about 75 percent of successful applicants matriculate.

The basic admissions requirements include graduation from secondary school, and completion of the Scholastic Aptitude Test (SAT) and three achievement tests (ACH) in subjects of the applicant's choice. An applicant's high school academic load should include four years of English, four years of mathematics, three years of science (with labs), three years of foreign language, and three years of history. Some training in music and art is also recommended. An applicant must have an interview, either on campus or with a local alumna or alumnus, and provide letters of recommendation from three teachers.

Such are the perfunctory requirements. Most applicants meet them easily. To actually get in, an applicant must have excellent grades and test scores: 98 percent or so of successful candidates rank in the top fifth of their high school class. Average SAT scores of those admitted are around 660 verbal, 700 mathematics, for a combined score of 1,360 out of a possible 1,600. (The national SAT average is 896 combined—422 verbal, 474 mathematics).

Still, straight A's and near perfect SAT scores alone do not guarantee admission to Harvard. If they did, it would be very difficult to decide between the fourteen thousand applicants, most of whom distinguished themselves academically in high school. An applicant must also meet other important criteria. Harvard's thirty-five member admissions committee looks for special talents—virtuosity on the flügelhorn, say, or a unique knack for computer programming. Extracurricular activities are important, and ethnic origin and geographic location play a role.

It has been said that Harvard is less interested in well-rounded students than in a well-rounded class: the budding poet with no aptitude for physics might get in, and so might the brilliant amateur chemist who has never read a novel that wasn't assigned in English class. A well-rounded class also means a broad geographic and ethnic distribution of students. In a recent freshman class, all fifty states and dozens of foreign countries were represented. Almost as many students now come from out west as do from New England. In terms of ethnicity, better than one student in four is of a minority, and about 45 percent of entering freshmen are women. Roughly 70 percent of all students receive some kind of financial aid. These figures represent a dramatic departure from the stereotypical Harvard student of years gone by: an affluent white male graduate of a New England prep school.

Since Harvard wants a diverse class profile, a geographic edge can sometimes make the difference between acceptance and rejection: a kid from Montana who is class valedictorian, star quarterback of the football team, and scored 1,400 on his SATs might get the nod over a kid from Massachusetts with equal accomplishments. This is because, in essence, the student from Montana is competing with other students from Montana, where the applicant pool is relatively small, while the Massachusetts student is competing against all other applicants from Massachusetts, which leads the fifty states in number of Harvard applicants. Conventional wisdom has it that the best way for parents to get their kids into Harvard is not to send them to an exclusive prep school in the East, but rather to move to some rural county in a state such as Montana.

The surest way of all would be for the parents to have graduated from Harvard themselves. About 40 percent of applications from children of Harvard alums are accepted; "legacies," as such students are called, are three times as likely to get into Harvard as unaffiliated applicants. Legacies constitute nearly 20 percent of Harvard's enrollment. Harvard admissions officials argue that the relatively high acceptance rate for sons and daughters of Harvard grads is only natural, since legacies have smart parents and usually come from privileged backgrounds where education is emphasized. The admissions people have also said that the legacy factor is only used to break a tie between equally qualified candidates.

Recently, however, the U.S. Department of Education's Office for Civil Rights decided to take a look at Harvard's classified admissions files. The OCR found that contrary to the university's claims, legacies were, on average, less qualified than their nonlegacy classmates. Based on an examination of admissions ratings of accepted students in a variety of categories, including academics, extracurriculars, personal qualities, and recommendations, the OCR concluded that "with the exception of the athletic rating, nonlegacies scored better than legacies in all categories of comparison."

The real reason so many legacies get into Harvard is financial: every year Harvard grads contribute millions of dollars to their alma mater. To shut the door on the children of wealthy alums might have a negative effect on their generosity. And that generosity, university administrators argue, feeds the scholarship fund that allows Harvard to admit eminently qualified students regardless of their ability to pay.

In the words of one Harvard official, "If you have a perfect SAT score, are first in your class and captain of an athletic team, you've got a good shot" at getting admitted. Add "child of Harvard-educated parents" to the equation, and you're a shoo-in.

?

How do chameleons change color?

THEIR LIPS CAN TURN YELLOW. Their strange little bodies fade from shoe-leather brown to grass green. The acclaimed herpetologist Angus Bellairs even reported in his 1970 two-volume work, *The Life of Reptiles,* that "if an object of distinctive shape is held against the flank of a [chameleon] for a couple of minutes, its outline will appear as a pale print on the darker skin."

Chameleon skin is full of cells called chromatophores, whose business is color. The bottom layer of skin is made of chromatophores filled with a dark brown pigment called melanin—the same pigment that colors human beings' skin. These particular cells are also called melanophores. The color cells above them vary, depending on the species of chameleon. Closer to the surface of the skin are cells that may contain yellow or red pigment; other cells have a substance called guanine, which is a reflector

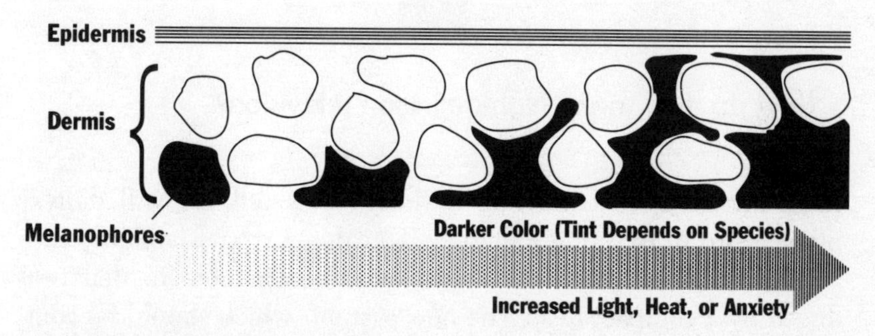

A chameleon can change color because of melanophore cells in its skin. When the chameleon is under stress, melanophores fill with pigment and surround all the other pigment cells.

and produces the effect of white color. When guanine is near melanin, it produces the color blue. Near yellow pigment, it produces green.

Color changes in chameleons seem to depend on the movement of melanin pigment into branches of the spider-shaped melanophore cells, which extend up toward the skin's surface. When the "legs" of the cells are full of melanin, the skin darkens. When they're empty, it lightens. The other pigment cells that the animal sports are either masked or exposed in the process, and wonderful colors result.

Although reports continue to say that chameleons change color to match their background, this doesn't always seem to be the case. Nervous system responses to light, heat, "emotions," danger and mating are highly correlated to color changes. Quite often a chameleon changes color and becomes *more* conspicuous. The melanophores of an anxious chameleon fill to the tips with melanin and the reptile washes dark brown. In a calm animal, on the other hand, melanin concentrates in the center of the melanophores and its color is pale.

?

How do they make high-definition television?

HIGH-DEFINITION TELEVISION, the experts tell us, will deliver pictures up to five times sharper than the current standard, with a larger field of vision and with sound comparable to that produced by a compact disk. The new system, which should be commercially available by the mid-1990s, will mark the first real change in television since color was introduced in 1954. The picture that appears on a standard television screen is transmitted as a series of images, each containing 525 horizontal lines of picture information, at a rate of thirty frames per second. All told, standard television pictures contain 212,520 dots of picture information, called pixels. High-definition television (HDTV) will achieve its revolutionary clarity by increasing to 1,125 the number of horizontal lines that compose each image and doubling the number of frames per second. HDTV pictures will contain five times as many pixels as current pictures, resulting in rich, seamless images. This is wonderful news for TV bugs, many of whom are probably ready to shell out the $2,500 necessary to buy a high-definition set.

But the change from traditional television to high definition involves more than tuning in a hypersensitive new television set: HDTV demands a brand-new way of broadcasting television signals, and the signals will be beyond the reach of ordinary television sets. HDTV carries too much information to fit into current broadcast channels. All those pixels can be squeezed into a standard channel about as readily as a camel can pass through the eye of a needle. Before the new sets begin appearing in stores, broadcasters must figure out how to broadcast high-definition signals without making obsolete the 160 million or so television sets now operating across the country.

To protect the consumer, the Federal Communications Commission (FCC) has ruled that HDTV will have to fit within current bands on the television spectrum. The sooner broadcasters figure out a way to comply with FCC requirements, the better, according to various electronics manufacturers who are dying to get their new high-definition sets on the market.

There are a variety of ways HDTV signals might be transmitted. One possibility would be for existing stations to send the extra signals necessary to complete a high-definition image through unoccupied channels on the television band. In local markets, to prevent interference, television stations are separated by vacant channel space. The FCC could assign each extant station an empty channel for the broadcast of HDTV signals. A station could broadcast one set of signals on ordinary VHF (channels 2 through 13) and send the additional picture information over a blank channel. Traditional TV sets would pick up the regular channel, while viewers with high-definition sets would pull in both sets of signals, which would be combined by their televisions into a single enhanced image.

Another possibility is direct-broadcast satellite. High-definition images could be beamed by satellite to homes equipped with satellite dishes. High-definition television sets would pick up both the new satellite transmissions and traditional broadcasts.

A third idea under discussion is to use the fiber-optic cables, currently being installed by telephone companies, to carry high-definition television signals. After buying new high-definition sets, viewers would subscribe to fiber-optic cable service, in much the same way as some traditional-TV-set viewers now buy cable TV packages.

When high-definition television eventually becomes available, the new hardware will have more in common with computers than with ordinary television sets. Traditional television works on an analog system: audio and visual elements are converted to undulating electric waves for transmission to television receivers (TV sets), which reconvert the waves into visible light rays (pictures) and audible sound. High-definition television will be digital. The new sets will convert analog waves into digital code, with strings of zeros and ones representing each bit of picture information—color, shade, placement. A microchip in the new sets will process the digital code, paying strict attention to each light element, to produce richly detailed pictures.

HDTV is not without its critics, some of whom contend that the enhanced pictures are barely discernible from those on old-fashioned TV. Even so, market analysts predict HDTV will be worth billions of dollars a year to the electronics industry; therein lies the real incentive. In the United States, where only one major company, Zenith, continues to make TVs, the financial motive is especially strong. The good news for American companies interested in regaining a toehold in the consumer electronics market is that current U.S. and Japanese HDTV standards are incompatible. The bad news is that Japan began HDTV broadcasts in 1989, giving it a big head start in perfecting the equipment.

?

How do they wrap Hershey's Kisses?

WHEN HERSHEY'S CHOCOLATE KISSES were introduced in 1907, just four years after Milton S. Hershey built his first chocolate factory, the bite-size candies were wrapped individually by hand. In 1921 the chocolate makers added the Kiss's identification plume, the little paper streamer printed "Hershey's Kisses" that sprouts from the top of the foil wrapping. The company made the name a registered trademark in 1923, and the candy's size, shape, wrap, and plume became registered trademarks the following year.

At about the same time that the Kiss was patented in the form that it has kept to this day, the company replaced its manual workers with a mechanized "single-channel wrapper." This handled the more elaborate wrapping and met increasing demand faster and more cheaply. The present-day "multichannel" Kiss wrapping machines are based on the concept and design of the original system, which was fed by one conveyor belt (single channel). Another five conveyor belts have been added per machine and other minor modifications made to speed up and otherwise improve performance. The current equipment now wraps approximately 33 million Kisses a day in the company's two factories in Hershey, Pennsylvania.

How, exactly, does it do this? First, the specially blended melted chocolate is squirted mechanically into uniform Kiss shapes on wide, continuously moving stainless steel belts that pass through a cooling tunnel for about eighteen minutes. The Kisses emerge in solid form, ready to be wrapped. The rows making up the broad array of Kisses are then separated. If a Kiss is properly aligned, standing upright on its base, it moves on with the others in single file onto narrow, steep-sided conveyor belts toward the wrapping machines. But if the Kiss has toppled over or is otherwise askew, a little mechanical finger pushes it off the line and it is set upright and mechanically recirculated onto the broad conveyor belt to wait its turn again.

A machine can simultaneously wrap the Kisses delivered to it by twelve such narrow conveyor belts, each of which passes over its own "wrapping chamber." Perforated squares of wrapping foil, still attached to one another on an unwinding roll but prepared like postage stamps to be torn off easily, are mechanically inserted through a side of a wrapping chamber and centered over its floor just in time to receive a Kiss brought onto it by the conveyor belt. When the Kiss is in position at the center of the piece of foil, a strip of tissue (the identification plume) is introduced mechanically from above, and at the same time a plunger pushes the Kiss down on the foil with sufficient pressure to tear off the section of foil without damaging foil or chocolate.

The wrapping chamber has a top and bottom section consisting of plastic-like bands of material that rotate in different directions and tighten against the Kiss, firmly twisting the foil around the chocolate. No sooner is the Kiss pressed down and the section of foil detached than the twisting mechanism is activated. This wraps the foil around the Kiss so that it captures and binds the printed streamer of tissue securely between foil and candy, allowing a jaunty plume to emerge at the top which is automatically cut to the proper length by a mechanical knife. Each step of the process is minutely calibrated to occur within a fraction of a second of those before and after it.

The instant the Kiss is wrapped, the conveyor belt moves it

out to be packaged with other Kisses for shipment, and brings along the next candidate. The entire wrapping process has taken about a second.

?

How do they teach a computer to recognize your voice?

IT WON'T BE LONG BEFORE you can carry on lengthy conversations with a computer, schedule business appointments by negotiating with a machine rather than a secretary, even instruct your VCR by voice when to record a certain show. IBM already has a computer in the works that can recognize twenty thousand words. How does it do it?

Basically, the computer "recognizes" your voice by converting its sound waves into digital signals. When you speak into a microphone, the computer compares your voice signals with a set of patterns it has stored in its system. The computer will show you what it thinks you're saying by instantly flashing up your words on a computer screen. You can then edit (by voice or keyboard), store, print, or transmit. In the labs at AT&T, for instance, researchers play video games, guiding a mouse through a maze simply by saying "left" or "right."

But before a computer can perform for you, it has to get used to your voice. (This type of computer will respond to only one voice and is dubbed "speaker dependent." Others, referred to as "speaker independent," will respond to any voice.) A speaker-dependent computer first listens for twenty minutes as you read a special document into a microphone. When you've finished, it will have stored two hundred sound patterns that characterize your voice, and when you use the system it will match them against your speech. A selection of words that roughly match those sound patterns is drawn from a twenty-thousand-word vocabulary. These words are then matched against a second model that has a vast data base of words commonly used in the office. The computer can reduce the number of possible candidates by determining which words are most likely to follow the two previous words. For example, the computer knows that *bows*—and not *boughs*—would follow the word *actor*. The system makes its final selection of the best word after it has determined that analysis of subsequent words (*to the audience*) won't affect its choice. Within a second or so, the word appears on the system's display screen.

This contextual ability enables the system to distinguish between words that sound alike but are different, such as *know* and *no*, *air* and *heir*, and *to*, *too*, and *two*. Punctuation can be added verbally by simply saying "period," "comma," and so on.

When IBM first started working on this system in 1986, the system occupied an entire room. Now it has been reduced to desk size and IBM feels it won't be long before it is used routinely in business offices. AT&T is experimenting with an automated directory system for the employees at Bell Labs. The caller simply picks up the phone, spells out the last name and first initial of the person he or she is trying to reach, and the call automatically goes through. Who knows? Before long you may be able to pick up the phone and just say, "Get me Mom, please."

?

How do they know who has won an election before the polls close?

IN 1980, the television networks caused a bit of a ruckus when they declared, before polls had closed in the West, that Ronald Reagan had won the presidential election. Soon after that announcement was made, with still more than an hour of voting time left on the West Coast, Jimmy Carter conceded defeat on national television. The scenario was repeated in 1984, when the networks correctly named Ronald Reagan the victor at eight-thirty Eastern Standard Time—voting rush hour in the West. In Congress, legislators from some western states raised the hue and cry, complaining that the practice of reporting results from the East before the polls had closed in the West disrupted voter turnout in the Pacific Time region. The truth is that the networks did not even have to wait for the polls to close in the East before knowing the results.

Journalists depend on exit polls, accurate to within a few percentage points, for their election numbers. Each network employs thousands of exit-pollsters to sample a fraction of the eighty million or so people who cast votes. The information collected by the pollsters, including party affiliation, candidate voted for, and reason for the vote, is fed into network computers, which prorate the data to arrive at state-by-state figures.

The exit poll was first used in 1967 by CBS in three separate state elections. During the 1972 presidential election, CBS conducted the first-ever national exit polls. And by the 1980 election

ABC, NBC, *The New York Times,* the *Los Angeles Times,* and the Associated Press had all joined CBS in carrying out exit polls. The big difference between an exit poll and pre-election tracking polls is that the exit poll measures a *fait accompli,* a cast vote, which is obviously more accurate than measuring *potential* votes before an election.

By 1985, the controversy surrounding exit polls—not whether they were accurate, but whether early forecasts deterred voters from turning out by creating in their minds the impression of a foregone conclusion—prompted the networks to sign a pledge to forswear "projections, characterizations, or any use of exit polling data that would indicate the winner of the election in a given state before the polls were closed." Some Western states had already taken matters into their own hands: Washington and Wyoming passed legislation prohibiting pollsters from interviewing voters within three hundred feet of a voting place.

None of the opposition to exit polls has prevented the media from conducting them; it merely stanched the flow of information. Whereas the networks once would have begun releasing figures the minute they became available, usually with the disclaimer, "Remember, this is not a result, just a trend," before going on to say, "but if the trend continues Hart looks all but unbeatable," the networks now wait until the polls close before stating what they have known for hours.

?

How do they train structural steelworkers to walk on unprotected beams five hundred feet in the air?

FIRST, WHAT GETS THEM UP there at all? "Youth is number one," says John Kelly, coordinator of the Joint Apprenticeship and Trainee Committee of Ironworkers Locals 40 and 361 in New York. "Adventure is number two. And it used to be a hand-me-down, father-to-son business. Or you'd be hanging around on the street like a bum and someone would say, 'Why don't you do this?' And when you were young you'd say, 'Gee, this is great work.' It's lovely up there. You're by yourself, or with your partner—it's like being up in a balloon. It's an interesting way to make a good buck. . . . And later you could look at the skyscrapers around town and say, 'I worked on that' or 'I helped build that.'"

It might be breeding, youth, good paychecks or pure guts that gets steelworkers and bridge painters comfortable at such great heights, but it isn't training. Apprentice school teaches safety and ironworking skills; the walking is up to the novice. An apprentice steelworker (more commonly called an ironworker) in New York City starts low—as low as twelve feet off the ground—working on the first floor of a building. Usually he's eighteen or twenty years old and cocky as they come. He looks forward to $27.50 an hour if he can advance to journeyman. If he's sure of his footing, he'll walk the five- or ten-inch-wide beams right off the bat. If he's shaky, he'll sit down, straddle the beam and monkey his way across. He sees experienced journeymen moving with ease around and above him. He notices that they watch the beam ahead of them and don't look farther (that is, lower) than their feet. Although most are extremely careful, some of the old pros will take unnecessary chances, actually jumping from beam to beam rather than taking the longer way around. When the apprentice thinks he's ready, he moves up and begins to work where high winds and dangling steel beams are a constant hazard.

Up until ten years ago, the unions lost up to six men a year on the job, some falling twenty or thirty stories to their deaths. Today, government standards and collective-bargaining agreements in New York City have considerably improved the working conditions, and fatalities have decreased. For example, every two floors of a building under construction must be solidly planked or decked over. If an ironworker happens to fall into, rather than out of the structure, he has only two-dozen feet to drop. Carelessness on the ground is more often the cause of accidents now: men have been hit by cars while setting up cones, and have fallen off ladders. For the painters, nets are hung under the bridges to stop a fall. You can imagine how grateful certain men have been for the union efforts to get the nets—plenty have experienced a fall that at one time would have meant sure death.

Despite such safety measures the work remains perilous. When you see a steelworker hammering a bolt into the outside skeleton beams at five hundred feet or more—standing there as if he were working on the bathroom closet—he *could* just as easily fall outside the building as inside. The bottom line is he's being careful.

?

How do they find heart or liver donors?

THE OBVIOUS OBSTACLE TO finding heart and liver donors is the singularity of these organs: a person has only one of each and cannot afford to part with either. A transplanted kidney, by comparison, can come from a live donor, who has two of them but needs only one to survive. The pool of heart and liver donors is necessarily small, consisting primarily of healthy people who die suddenly in accidents.

To simplify the search for transplantable organs, the National Organ Transplant Act of 1984 established the United Network for Organ Sharing (UNOS). UNOS maintains a data bank of available organs and matches donors and recipients according to location, compatibility and urgency of need. By law, all transplant centers must be tapped into the network. The law also makes provisions for acquiring organs: all hospitals receiving Medicare funds are required to ask the next of kin of brain-dead patients whether they want the dead relative's organs donated. Forty-two states provide organ-donor check-off boxes on their driver's licenses, but such identification is construed merely as a willingness to donate, and the final decision falls to the deceased person's relatives.

UNOS maintains local, regional, and national computerized lists of patients awaiting transplants; each patient is ranked according to need and the length of time he has been waiting. People waiting for new livers, for instance, are rated from 0 to 4: status 0 patients are considered temporarily unsuitable for transplants; status 1 patients are at home and functioning; status 2 patients require continuous medical care; status 3 denotes required hospitalization; and status 4 signifies acute and chronic liver failure. When a liver becomes available for transplantation, the computer scans its lists for compatible recipients. Final allocation is made based on compatibility, urgency and geography: local status 4 receives top priority, followed by all other local patients; next come regional status 4 patients, then all other regional patients; finally national status 4, and then all other national patients. Local and regional patients have priority over national patients because the transplant must occur as quickly as possible after the organ is removed from the donor.

Ideally, the UNOS system works like this: a person dies, needed organs healthy and intact. The hospital transplant coordinator is called in to talk to the family about organ donation. The family acquiesces. The organs are examined and the vital information—blood type, tissue type, size—is plugged into the UNOS data bank. The computer, beginning with the local list, scans the patients awaiting transplants. If no match is made locally, the computer jumps to the regional list, and finally to the national list. In 1989, organs were found for slightly better than half of the almost 23,000 people awaiting transplants.

It was in 1989 that surgeons performed the first liver transplant involving a living donor. The procedure allows doctors to replace a child's ailing liver with a piece of a liver cut from a healthy adult—for compatibility reasons, usually a parent. So far, surgeons have performed the operation only on children, because they are small and can survive without a full-sized liver. The technique is possible because the liver, alone among human organs, is capable of regenerating: what remains of the donor's organ will return to normal size after a few months, while the recipient's organ will grow as the child does. The operation could make it much easier for children to receive life-saving transplants—more than seven hundred require them each year—but the procedure is potentially very dangerous for the donor, leading some doctors to speculate that cadaver donations will always remain preferable.

Once an organ has been found and transplanted, the recipient faces the hurdle of organ rejection, which occurs when the recipient's body attacks the new organ as a foreign object. Rejection is very common—it occurs in all cases of heart transplantation—even when the donor and recipient have been closely matched by blood and tissue types. The only way to treat rejection is by suppressing the recipient's immune system with steroids, a strategy that can make even a common cold fatal.

?

How do they get the cork into a bottle of champagne?

TWO SECRETS APPLY: first, the cork was soaked and thereby softened before being driven into the bottle, and second, the cork was originally a bit slimmer than the one you see fly across the room on a festive occasion—it has expanded over time because of the pressure of carbon dioxide within the bottle.

True champagne is a sparkling wine from the Champagne region of France. In contrast to still wines, which are fermented once, champagne is fermented twice, once in a cask or tank, a second time right in the bottle. Fermentation occurs when yeasts act upon sugar in the wine, converting it to alcohol and carbon dioxide. One of the first to make a study of this double fermentation was a monk by the name of Dom Pierre Pérignon, who became cellar master of the abbey of Hautvillers in 1670. He toyed with the sugar content of the wine and so began to learn how to control the bubbliness of the stuff. He also garnered wines from different fields and blended them in order to produce the same taste year after year, a practice continued to this day. While at

Hautvillers Dom Pierre encountered some Spanish pilgrims whose gourds were sealed with cork. Until that time he had been stopping his bottles with wooden plugs wrapped in oiled hemp. He quickly saw the advantages of the more elastic substance, which expanded in the bottle, sealed it better, and reduced leakage and explosions.

Cork trees grow in the Iberian peninsula, southern France and southern Italy, Corsica, Sardinia, and North Africa. The best corks—those suitable for capping a fine champagne—come from Spain and Portugal, from cork trees at least fifty years old. After a lengthy process of drying, boiling, and cutting, the corks are milled with emery to produce a smooth finish and sometimes treated with paraffin to help them slide more easily into the bottle. Champagne corks are forty-seven millimeters long and thirty to forty-eight millimeters in diameter, substantially larger than corks for still wines. The substance is ideal for its purpose, being durable and impervious to liquid. And when it is cut, some of the cells that are divided develop the characteristic of adhering well to a slippery surface.

After a bottle of champagne has fermented the second time and been left to ripen in a cool cellar for a number of years, it is skillfully turned each day and eventually left standing on its head. Any sediment in the wine falls down to the cork. Then the bottle is ready for *degorgement,* in which a highly trained worker pries off the first cork, allowing sediment and as little wine as possible to fly out of the bottle. Today the neck of the bottle is often frozen first to facilitate the procedure and reduce waste. When the bottle is opened, a frozen lump of sediment and a little wine is shot out by the buildup of pressure within. More wine is then added to replace what was lost, usually along with a liqueur of aged champagne and sugar. At last the bottle is ready for the final cork. Corking machines have clamps that compress the cork, making it smaller than the neck of the bottle, and a vertical piston that drives the cork into place. It will expand in the bottle where pressure is five to six times that of the outside atmosphere. Finally, other machines add metal caps secured with twisted wire.

?

How do they make Hostess Twinkies stay fresh for years and years?

PERHAPS YOU'VE HEARD health nuts remark that Twinkies, the ultimate junk food, are so loaded with chemicals they can sit on the shelf for twenty years and the consumer wouldn't know the difference. Not so. "That's a popular misconception we're not too happy about," says Patrick Farrell of Continental Baking Company, producers of the familiar creme-filled snack. "The shelf life is actually seven days," says Farrell. "Then they go to thrift stores. After fourteen days, you wouldn't want to buy them. They taste like cardboard." Are any preservatives used? "Just flour extenders," claims Farrell, "and the kind of preservatives used in any commercially baked bread."

Despite the protests of health fans who would choose a more nutritious snack, Americans apparently don't leave Twinkies on the shelf for long. They consume some 500 million of them a year. It takes forty thousand miles of cellophane to wrap those Twinkies—enough plastic to encircle the globe one and a half times at the equator.

?

How does the SEC know when someone's doing inside stock trading?

INSIDE TRADING, the illicit profiteering by brokers on stock sales based on advance information about companies, is illegal because it gives the guilty trader an unfair advantage over his competitors; it can also lead to a fiscally damaging betrayal of trust. A lawyer, for instance, retained by Huge Company to assist in its secret buyout of Lucre, Inc., could stand to make a quick buck by investing heavily in Lucre before the buyout goes public, at which time the price of Lucre stock would inevitably rise. Through his work for Huge Company the lawyer would have gained an advantage on the stock-trading public. By capitalizing on his inside information, the lawyer would also hurt his client, Huge Company, by driving up the price of Lucre Inc. stock, thus making the buyout more expensive. The shifty attorney could further increase his windfall by selling his scoop to an amoral broker, who would profit by investing heavily in Lucre Inc., just before news of the buyout was released to the public.

The main way in which the Securities and Exchange Commission (SEC) learns of possible inside trading is by monitoring stock sales. By watching Wall Street trading, the agency might notice a jump in the stock of a target company just before a takeover bid is announced. If, over the course of months, it becomes apparent that a certain broker has repeatedly capitalized on corporate takeovers by the early purchase of target company stock, the successful broker might become an inside trading suspect. Proving the illegality of a trade is not so easy: the broker could always argue that his trades were guided by market research and accompanied by luck.

The SEC only really knows that someone is engaging in inside trading when an informant tips it off. Until then, the agency may possess suspicions, but not definite knowledge. Once tipped off, the SEC examines the record; if it finds a disturbing pattern of trading, it serves subpoenas to every person remotely connected with the fishy trades. Usually a case can be pieced together from the testimony of the subpoenaed dealers; then the agency can present its evidence to the alleged inside trader in hopes of squeezing out a confession.

The biggest inside trading scandal in the history of Wall Street began to unravel in 1985, when the SEC received an anonymous tip about the fantastically acute transactions of Dennis Levine, a thirty-two-year-old mergers and acquisitions specialist with Drexel Burnham Lambert. The agency decided to examine the young investment banker's record. With the help of the New York Stock Exchange, which maintains a computer system to monitor stock trades, the SEC found fifty-four incidents of prescient trading by Levine. Apparently his $3 million in salary and bonuses was not enough. The investigators began contacting Levine's more devoted clients, companies that had made a lot of money through their dealings with him. By May of 1986 the SEC had unearthed enough frightened investors to corroborate the case against Levine. Levine, realizing he had been caught with his hand in the cookie jar, made a bid for leniency by cooperating with the investigation. Then the dominoes began to fall.

Based on testimony from Levine, the SEC fingered Ivan Boesky, the biggest arbitrageur in town. Boesky was eventually sentenced to three years in jail and ordered to pay penalties totaling $100 million, but not before he agreed to cooperate with the SEC by wearing a secret wiretap during his illicit dealings. His conversations proved damning; after weeks of taping the SEC began preparing cases against Martin Siegel, a Wall Street mergers expert, and Michael Milken, Drexel's infamous junk bond king.

Because it would be very difficult to convict an inside trader solely on the basis of suspicious trading patterns, the SEC relies a great deal on the kindness of strangers. As Gary Lynch, the SEC's enforcement director put it, "There's nothing more important than a phone call from someone with helpful information."

?

How do they decide what goes on the cover of *People* magazine?

THERE'S AN OLD FORMULA at *People* called the Law of Covers: "Young is better than old," it goes. "Pretty is better than ugly. Rich is better than poor. TV is better than music. Music is better than movies. Movies is better than sports. And anything is better than politics."

The Law was conceived by *People*'s founding editor, Dick Stolley, back in 1974 when the pop-culture magazine was launched, and today it's still a pretty good formula for predicting newsstand sales. The people on *People*'s covers are largely responsible for the weekly's phenomenal success on the newsstands, where slightly more than half the readers buy their issues. It's as simple as this: if you see someone on the cover you want to read about, you'll buy the magazine. But it's also as complex as this: Why didn't you buy the issue with Roseanne Arnold on the cover? And why *did* you buy the one (that seemed riskier) featuring New Kids on the Block? Anyway, it all evens out. In 1991, *People*'s weekly newsstand sales averaged 1.7 million copies; the only weekly boasting higher numbers was *TV Guide*.

Stolley's Law needs two amendments to bring it completely up to date. Tributes to stars who've died suddenly or catastrophically sell more magazines than any covers in the history of *People*. And royalty, particularly of the British persuasion, came into vogue with *People* readers in the 1980s with the wedding of Prince Charles and Lady Diana. In fact, Diana is *People*'s favorite "cover girl," having appeared on the cover forty-four times as of 1991. These are the other top favorites: Liz Taylor, John Travolta, Sarah "Fergie" Ferguson, Jackie Onassis, Farrah Fawcett, Cher, Madonna, Princess Caroline, Bruce Springsteen, Michael Jackson, Olivia Newton-John, Jane Fonda and Brooke Shields. And here's the list of *People* covers that ranked highest in newsstand sales as of this writing:

John Lennon: A Tribute (December 22, 1980)
Princess Grace: A Tribute (September 27, 1982)
Royal Wedding of Prince Andrew and Sarah Ferguson (August 4, 1986)
Princess Diana: "Oh Boy!" (the birth of Prince William; July 5, 1982)
Royal Wedding of Prince Charles and Lady Diana (August 3, 1981)
Karen Carpenter's Struggle with Anorexia Nervosa: A Tribute (February 21, 1983)
Olivia Newton-John (July 31, 1978)
Brooke Shields and Chris Atkins: *The Blue Lagoon* (August 11, 1980)
Priscilla Presley Talks about Elvis (December 4, 1978)
Farrah Fawcett—Why She Left Lee Majors (August 20, 1979)

While Stolley's Law is a nutshell formula for success, *People*'s cover subjects can't always be pretty, rich TV stars; in fact, the cover occasionally features such political personalities as Ted Kennedy and the Reagans. A more general rule of thumb calls for cover stories that grab you because they're "what you and your friends are talking about on Saturday night," says Assistant Managing Editor Susan Toepfer. To find out just what that is, *People* sometimes invites readers at random from the streets of New York City for a "cover lab," says Toepfer. These readers are asked to vote for potential covers that appeal to them most and to answer specific questions about which headlines and photographs they like. "But you can't take that as gospel," Toepfer remarks. "What they say and what they do are not necessarily the same." So most of the time, the cover decision is made in a weekly meeting of *People*'s top editors, a group of about fifteen. They plan enough cover stories for two months ahead and pick two or three cover subjects for the next week. By the following Tuesday, or Wednesday at the latest, the cover choice is narrowed to one, and that one is sent to press.

Unless, that is, something happens between the meeting and press day: war breaks out in the Persian Gulf, for example, or a princess's husband dies in a tragic boating accident. Toepfer points out that *People* is news-oriented, with the kind of staff that can research, write, edit, photograph and fact-check a story in a few hours. For example, when Lucille Ball died on Wednesday, April 26, 1989, the editors decided to hold the cover they had planned for the next issue (which would be dated May 8 but was due at the newsstands on May 1). Then at least one senior editor, two writers, five correspondents, and a crew of photo and editorial researchers put together an eleven-page tribute to Lucy, in what was probably record time even for *People:* about two and a half hours. "Lucille Ball: 1911–1989" was that year's best-selling cover.

What about covers that don't sell? "We're often stunned by a cover that doesn't sell," Toepfer says. Here, in order of bad to all-time worst, are the all-time lowest newsstand sells:

Harry Hamlin: Sexiest Man Alive (March 30, 1987)
How to Make Your Kid a Star (November 12, 1984)
Ted Danson (May 11, 1987)
Dustin Hoffman and Warren Beatty (May 25, 1987)
Broadcast News (February 1, 1988)

Jacqueline Bisset and Alexander Godunov (April 1, 1985)
Avenging Sergeant Kenneth Ford (April 28, 1986)
Jay Leno (November 30, 1987)
Michael Caine (May 4, 1987)
American Hostages in Lebanon (July 18, 1988)

?

How do they put the smell into scratch-and-sniff advertising?

IN YET ANOTHER clever ploy to attract customers, Madison Avenue invented the scratch-and-sniff ad. Used mostly by the cosmetics industry to sell perfumes and colognes, the ads are inserted into magazines, so you can get an idea of what the product smells like before buying it. Scratch-and-sniff strips are composed of globules of oil-based scent coated with gum arabic, which are suspended in glue for printing. When you scratch the strip, you burst the gum that hold the globules and the smell escapes.

To create the globules, the laboratory takes an oily essence of any odor desired and adds it to gum arabic in a "jacketed vessel," a steel container of water in which a glass vessel floats. (The double layers keep the mixture from getting too hot so the fragrance won't break down and escape into the air.) Attached to the side of the vessel is a projecting arm with an electric motor on top and a hanging paddle below for stirring. When the paddle moves, it beats the mixture into an emulsion: the scent breaks up into droplets and the gum arabic surrounds them and coats them. The speed of the stirring and the size of the vessel determine the size of the capsules created. "When you go a little slower than an eggbeater you get capsules about fifty microns in size—each micron being one millionth of a meter," explains Jim Conklin, a chemist at Lawson Mardon, Inc., one of the packaging and printing firms associated with the popularization of scratch-and-sniff advertising in the early seventies.

After about two hours of stirring, the lab technicians place the mixture on mesh screen, allowing the unbonded materials to drain away, and wind up with a moist cake of scent about the consistency of coarse sand. The technicians then transfer the cake to a new vessel, gently break it up with a spatula, and add water and an ink-paste with an alcohol base. Then the globule mixture is ready to print.

Just like ink, it lies in a pan underneath the press and a roller passes over it, dipping in a little rubber pad which prints a specified amount on each sheet of paper that goes by. Ordinarily, the mixture is applied after all the rest of the printing has been done; then the paper is ready to be whisked through a dry oven on a mechanical belt at about fifteen miles per hour—often the last step in ordinary four-color printing. Finally the scratch-and-sniff strip is ready: the rich globules lie waiting for a finger to release their cache of scent to the air.

?

How does sunscreen screen out the sun?

AS THE OZONE LAYER wears thinner (see "How do they measure the ozone layer?"), the risks of skin cancer from sun exposure increase. About 90 percent of all skin cancers arise from exposure to the deceptively kind, warm rays of the sun—hence the advisability of sunscreen on a sizzling day at the beach.

Sunscreens are classified as drugs and as such are regulated by the Food and Drug Administration (FDA). Manufacturers use FDA-approved chemicals—oxybenzone, padimate-O, octylp-methoxy-cinnamate, and others—in different amounts and combinations, each arriving at its own unique and well-guarded formula. What these chemicals do is absorb or reflect radiation in the ultraviolet spectrum; older products like zinc oxide, on the other hand, are opaque and mechanically block radiation, including visible light. When the sun's rays strike the chemical molecules in sunscreen a transformation occurs: energy in the form of light is converted into a different kind of energy—heat, which is given off into the air. This phenomenon, in photochemical terms, is *absorbance.*

Ultraviolet radiation is characterized by the length of its waves, which are measured in nanometers, one nanometer being one millionth of a millimeter. At one end of the range of wavelengths called ultraviolet are the shorter wavelengths of 280 to 320 nanometers. These high-energy rays are called ultraviolet B (UVB), and they are responsible for causing visible sunburn. Longer ultraviolet rays, at 320 to 400 nanometers, are known as ultraviolet A (UVA). Scientists have recently found that they penetrate the skin more deeply than UVB and, though leaving no evidence of sunburn, do cause skin damage. Most sunscreens are designed to block UVB, and the sun protection factor (SPF) listed on the product refers to its effectiveness against UVB. However, according to Patricia Agin of Schering Plough HealthCare Products, any sunscreen with an SPF of over 10 will also provide some protection against UVA.

?

How does bleach get clothes white?

MOST LAUNDRY BLEACHES, including the household wizard Clorox bleach, are oxidizing agents. In the washing machine they release free-roving molecules of sodium hypochlorite or peroxide. The color of a stain or spot is made up of a group of atoms and molecules linked together by a pattern of double and single bonds. The oxidizing agent tears into those bonds, destroying the bond pattern and fading the color or changing it completely to white. The stain is still there, albeit invisible, until detergent and the agitation of the machine lift most of it off.

Fabric colors are also made up of bonds, so that if you add bleach to the wrong kinds of wash loads—clothes that aren't colorfast—you'll notice that the colors you liked might also become invisible.

?

How do they get rid of radioactive nuclear waste?

THE ANSWER IS . . . not very well. Ridding the earth of nuclear waste that continues to emit radioactivity is a formidable, if not impossible, task, and experts disagree over how it should be packaged and where it should be put.

There are two general types of nuclear waste—low-level and high-level. Low-level waste is composed mainly of the by-products of nuclear reactors, such as contaminated work gloves, tools, and irradiated reactor components. Such waste need not be shielded from personnel and does not require heat-removing equipment. Often it is filtered (much like coffee), reduced to sludge and burned in incinerators. The resultant gases contain very little radioactivity, but the ashes must be treated as low-level waste and buried.

Some low-level waste is compacted in order to reduce its volume and produce a more stable structure for disposal; some is solidified by mixing it with concrete or asphalt to resist attack by water and make the waste less subject to leaching or seeping out.

Next, the low-level waste is packed in reinforced plastic-and-concrete or steel containers and buried twenty-five feet or more beneath the surface. Much of this burial (called shallow land burial) is currently done on-site, but efforts are under way to find permanent burial places that are not subject to volcanoes, earthquakes, landslides or excessive weather and are far from commercial and housing sites. Moreover, such areas must have a water table deep enough to prevent immersion of wastes, and contain no springs to bring contamination to the surface. Of course, these areas must never be mined.

Alternative methods of disposal being considered for low-level waste include below-ground vaults with a drainage channel (to prevent water gathering) and a clay-topped concrete roof. Above-ground concrete vaults, another option, would be more accessible and the waste more readily retrievable for monitoring purposes, but might be eroded by the weather, particularly by acid rain. Some experts favor sinking deep into the ground a shaft with radiating corridors at the bottom for storage.

High-level waste consists mostly of liquid from defense production and of used fuel rods (produced from uranium) that have powered nuclear reactors for three to four years and continue to be radioactive long after their fuel is spent. These materials must be immobilized before being placed underground. Spent fuel rods, which are hot and highly radioactive, are stored underwater in large pools or stacked in racks at the reactor site until they cool down sufficiently to be packed into containers and buried. Liquid waste must be mixed with some material that will resist water or any other solution that may be underground. This material must also be immune to heat and radiation from fission decay. Most liquid waste is combined with borosilicate glass containing glass-forming chemicals. The mixture is then heated to 1,150 degrees Celsius, which not only destroys organic matter but also dissolves inorganics into the glass.

Interment is a problem because high-level waste remains radioactive for centuries (eighty thousand years for plutonium). So far, burial in salt has been the preferred choice because it is abundant, far from earthquake zones and inexpensive to excavate. If a hole could be made sufficiently deep—say six to ten miles below the level of moving groundwater—other types of terrain could be considered, but little is known of geologic activity at that depth. Scientists are considering dropping high-level waste onto rocks a mile beneath the earth; the rocks would melt from radioactive decay, mix with the waste, solidify and finally, cool down after a thousand years. Another remarkable solution still in the testing stages is hydrofracturing: first, water is forced into rocks such as shale, causing the layers to separate, then the waste mixed with cement or clay is pumped into them and left to harden.

Scientists have dismissed the idea of shooting waste into the sun or space and letting it orbit, because of the likelihood of accidents. Allowing decay heat to melt through either polar ice cap, which would then form a freeze seal above it, has also been rejected because of difficult weather conditions and the problem of coordinating different countries involved in such a project.

Solutions for the safe disposal of nuclear waste should be found before we permit more of it to accumulate. We have not yet disposed of waste produced from World War II. Although the Nuclear Regulatory Commission claims that the hazard from nuclear waste which seeped out of single-walled steel containers buried in shallow trenches was not a widespread threat, it nevertheless contaminated the surrounding earth and air and has caused certain areas, in Washington State and New Mexico for example, to be designated "sacrifice" zones.

The sad fact is that although billions have been spent in the last forty years on producing nuclear defense weapons, only $300 million has gone toward researching and testing what to do with the deadly waste that is left behind.

?

How do they keep Coca-Cola drinkable in polar regions?

IN VERY COLD CLIMATES, the easiest way to keep Coca-Cola from freezing is to store it inside, in a heated shop or home. Exposed to the elements for an extended time, Coke would freeze, causing an indefinite postponement of the pause that refreshes. Actually, before the soda could freeze, the can itself would burst. The aluminum cans are structurally incapable of handling the change in pressure brought on when the liquid within them freezes. So, even though Coke is best when served cold (Coke officials say 40.1 degrees Fahrenheit or below is ideal), it won't keep well in a snowbank. It can, however, be stored in an outdoor vending machine, even in subfreezing temperatures.

The trick to outdoor vending machines is to keep them warm enough to prevent the beverage from freezing, but cool enough that the Coke isn't heated. Warm Coke is even less appealing than frozen Coke. Large heating units, therefore, are not desirable. Instead, cold-climate Coke machines are equipped with a sixty-watt light bulb and a fan. The light bulb, situated at the top

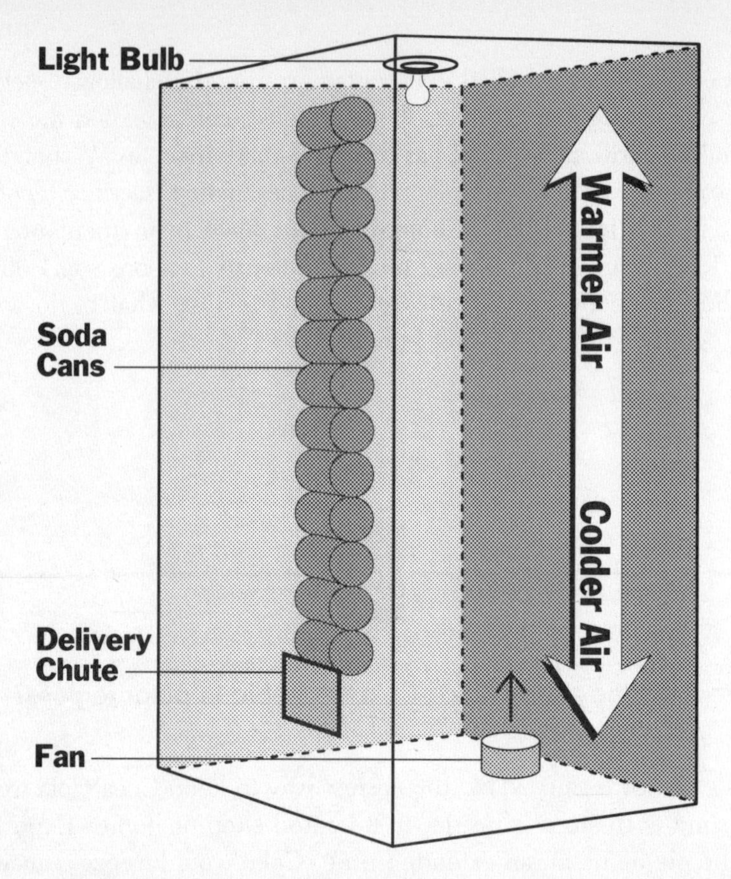

of the machine, generates heat; the fan, situated at the bottom of the machine, circulates cold air from the bottom of the machine to the warm area at the top.

Cans of Coke are stacked lengthwise in a chute inside the machine. The circulation of air within the machine ensures that all the cans are exposed to warm air; because hot air naturally rises, and because the source of heat—the light bulb—is at the top of the machine, the cans at the top of the chute stay warmest. When a person deposits money in the machine, the bottommost can in the chute is dispensed and all the other cans drop down one position. With each drop in position, the cans experience a corresponding drop in temperature, so that by the time a can reaches the bottom of the chute and is dispensed, it has achieved perfect drinking temperature.

?

How do they know if a runner false-starts in an Olympic sprint?

A ONE-HUNDRED-METER RACE lasts about ten seconds, so the difference between first and last place may be mere tenths of a second. The incentive is high to get out of the starting blocks as fast as possible, but if a sprinter false-starts more than once in a race, he is automatically disqualified. False starts just make the other runners skittish.

The starters are responsible for judging whether a runner moved before the gun was fired, and for the most part they do that with the naked eye. Like an offsides referee in a football game, or an umpire in baseball, they've watched so many races they can detect a runner's slightest motion in the blocks. The best of them can keep the runners calm and release them smoothly out of the tense "set" position and into the race. Basically, starters are not "trying to fool the runners," says Olympic sprinting coach Mel Rosen.

Starters for the Olympics and a few other high-level track meets do employ a clock and computer as a backup to their well-trained eyes. The clock and computer are wired to pressure-sensitive pads in the starting blocks and to the starter's pistol. If a runner pushes off the pads before the gun is fired (or too soon afterwards to be within possible human reaction time), the starter hears a tone and calls a false start. Usually the starter uses the information only to double-check the lane number in which a false start occurred.

Human reaction time is a bit tricky to define categorically, especially when Olympic athletes are the humans reacting. The electric starting blocks at the second World Championships in Athletics in Rome in 1987 were set for a reaction time of 0.120 of a second, or 120 milliseconds. Canadian sprinter Ben Johnson left the blocks 109 milliseconds after the gun was fired but before the set human-reaction time. The starter did not call a false start, and film footage from the race confirmed the starter's judgment. Johnson just happens to react fast, *and* run fast. He placed first in the race. Coach Rosen noted, however, that Johnson uses an unorthodox start: he throws forward his hands, under which, of course, there are no pressure-sensitive starting blocks.

?

How do crocodiles clean their teeth?

THE EGYPTIAN PLOVER (*Pluvianus aegyptius*) has long been called the crocodile bird for its symbiotic relationship to this most ferocious of reptiles. Sitting coolly in the gaping mouth of a basking crocodile, the bird picks leeches, flukes and tsetse flies from the animal's gums to eat.

In nearly all other circumstances the crocodile is an extremely dangerous creature, singularly responsible for dozens of human lives lost every year in Africa. But it seems to have an instinctive tolerance for the plover.

Although the bird is a definite asset to the crocodile, its value as a hygienist is a little overrated. Crocodiles have widely spaced teeth, and rarely chew their prey but just bolt it down—so there isn't much call for flossing. Also, the crocodile sheds worn, decayed and broken teeth regularly throughout its fifty- to one-hundred-year life and replaces them with fresh ones. Humans, of course, "shed" their teeth only once.

?

How do lawyers get paid when a company files for bankruptcy?

UNDER THE FEDERAL BANKRUPTCY CODE, administrative costs of running a bankruptcy are given priority over debt owed to creditors. Therefore, both the counsel for the debtors and the official committees representing the creditors are ensured payment because technically they constitute an administrative expense of running the case.

A lawyer cannot represent a bankrupt entity. So, when a company voluntarily files for bankruptcy, attorneys are paid for all fees and expenses up until the time that court proceedings begin. Thus it is not unusual for attorneys to be compensated in large sums on the eve of the commencement of the case. When a company is in distress and may be forced into involuntary bankruptcy, lawyers make sure that they are paid routinely and promptly in the event such action occurs. In other words, they don't want to be stuck with a huge unpaid bill if the company's assets cannot cover its liabilities.

When the judicial procedure for filing for bankruptcy begins, so does a whole new tab on the behalf of the legal counsel. Lawyers for the bankrupt entity and the official committee of creditors then make applications to the court to be retained. At this point, the counsel for the official committee of creditors may object to fees and expenses proposed by the debtor's attorneys. Once rates and expenses are settled, counsel proceeds with the work and submits quarterly bills for review by the court. Once again this application is subject to objection from a creditor or other party of interest. If the court so decides, a hearing is held between lawyers for the debtor and those for the official committee of creditors. Often, creditors are very precise in their protestations; they will question, for instance, why three lawyers participated in a conference call rather than two, or why the debtor is being charged twelve cents for each photocopy rather than ten cents. If the court agrees, rates and fees are adjusted accordingly. Local judges in some parts of the country will not allow New York lawyers to be paid their hometown rates. For this reason many high-paid attorneys reduce their fees by as much as one third to ensure court and creditor approval.

After the court approves these interim payments, a percentage is usually withheld until the case is finished. In New York's Southern District, for instance, that percentage is typically 25 percent, to be paid from the debtor's estate when the case closes.

?

How do they decide how much to pay the queen of England?

CALCULATING HOW MUCH TO PAY the richest woman (and fourth richest person) in the world is a job which may seem akin to deciding how much sand to take to the beach, but HRH Elizabeth II, and the whole Windsor clan, take the matter quite seriously. In fact, despite complaints and resistance from Labour party MPs, Her Majesty has asked for a raise on more than one occasion—and her requests have been met.

While the queen's cash flow comes from various sources—rents, stocks, foreign investment—her actual *salary,* which in 1989 was $7.9 million, is drawn from the public treasury called the Civil List. In addition to that salary, she is reimbursed from the Civil List for her expenses incurred during her duties and for the cost of running Buckingham Palace, Windsor Castle and Holyrood House. A few expenses that year included, among other things, $350,000 for garden parties, $200,000 for care of her ceremonial horses, $98,000 for flowers and about $6.2 million in staff salaries. However, no one is getting rich working for the queen: a footman will only make $12,000 in a year. She also subsidizes relatives like the duke of Kent and Princess Alexandra, first cousins not provided for by Parliament.

Increases in the Civil List are decided by the House of Commons, based on information regarding her expenses assessed and provided by a select parliamentary committee headed by the chancellor of the Exchequer. Her first raise was not granted until 1971, nineteen years after she ascended the throne. (Her starting salary and expenses were $1,187,500. Her first raise brought the Civil List up to $2,450,000.)

In 1989 the Civil List totaled $10.5 million. In July of the following year a vote in Commons increased it to $20.3 million, a figure which will hold for ten years. In arguing for her raise, sympathetic MPs will often point out that the queen performs her state duties with skill and dignity. In 1989 those duties included 147 official visits and opening ceremonies, 73 banquets and receptions, 18 working days abroad and 268 other audiences.

Expenses not covered by the Civil List are often provided for by the government in other ways. For example, the Ministry of Defence paid $33.2 million to refit the royal yacht *Britannia;* physical upkeep of the royal palaces is paid for by the Department of the Environment; and royal postage, stationery and telephone service is free. Costs of maintaining the royal trains and aircraft are also deferred directly to Parliament.

The Civil List was created in 1760 when George III agreed to turn over all of England's crown properties to the state, with the exception of the 55,000-acre Duchy of Lancaster, and the 128,047-acre Duchy of Cornwall. Rent collected on the former property is reserved for the reigning monarch. (The latter property is the domain of the heir apparent—in this case, the queen's eldest son, Charles, prince of Wales.) That money, the "privy purse," is intended to pay for the queen's personal expenditures, such as clothing and the upkeep of Sandringham and Balmoral castles, which she owns outright. Occasionally, when the costs of running her residences and public entertaining exceed her official grant from the Civil List, the queen will dip into these funds to make up the deficit. It is at that point that she'll put in for a raise.

The best part of all this? In Britain, where the top tax rate is 40 percent, all of the queen's income is tax-free. Of course, this may change in the future. She does, however, voluntarily hand back to the state some of her salary. Even with that concession, her total worth in 1991 was estimated by *Harpers and Queen* to be $13 billion.

?

How do they raise a sunken ship?

IT ALL DEPENDS ON the ship and its location in the water. If the main deck of a stranded vessel is above water, salvagers patch the holes that caused the ship to sink and pump out the water. The lightened ship will then refloat. If the main deck is submerged, but not deeply, it is possible to build a waterproof wall, called a cofferdam, that extends around the deck and above the surface. Once the cofferdam is in place, the water in the cofferdam is pumped out and the vessel is patched and refloated. Another

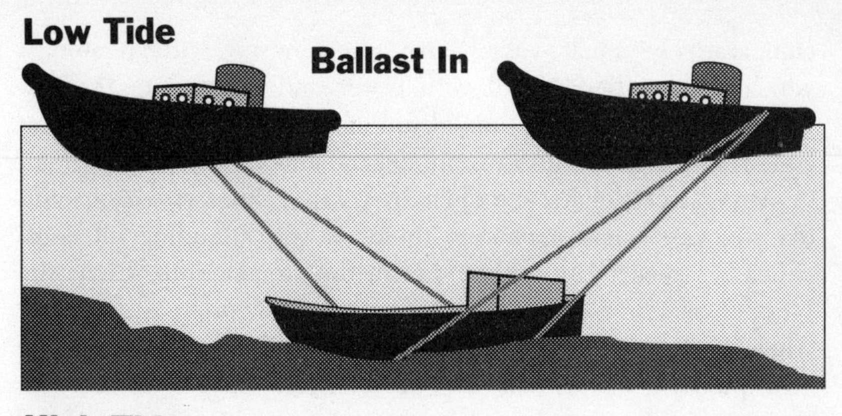

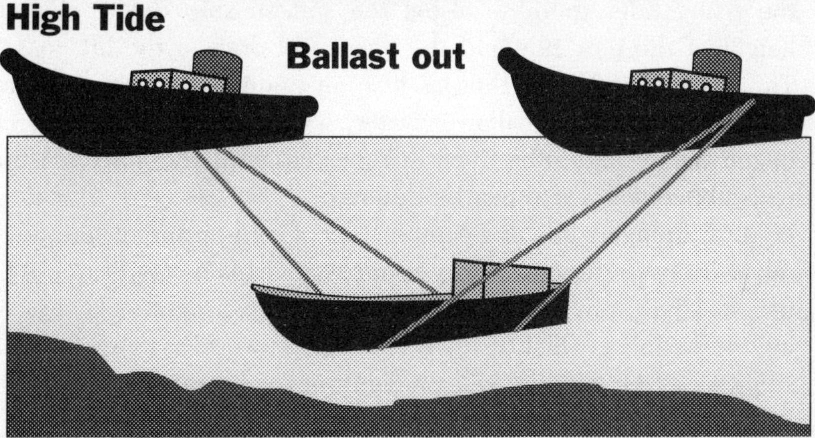

Cables holding a sunken ship to the rescue craft are tightened at low tide, when the distance between the rescuers and the wreck is smallest. All three vessels are lifted on the rising tide. The ships are raised further when the rescue craft discharge ballast, increasing their height in the water.

method, known as foam-in-salvage, works by displacing water with urethane foam. A diver armed with a massive hose fills the ship with a mixture of liquid polyol, fluorocarbon gas, and liquid isocyanate, which combine to form a quick-hardening, lightweight foam. The foam forces water out of the ship, and the buoyancy of the foam raises the ship to the surface. Two similar methods call for expanded plastic beads and hollow polystyrene balls about the size of pumpkins instead of foam.

Ships weighing less than a thousand tons can be hauled to the surface by floating cranes and then pumped out. Barges and other comparatively small vessels are often raised with pontoons, which, having been filled with water and chained to the distressed craft, are then pumped full of compressed air. The giant balloons float to the surface, pulling the attached ship with them.

Divers recover heavier ships by passing thick wire cables under the wreck and anchoring them on two attending lift craft, which are moored on either side of the sunken ship. First, the lift craft take on ballast water to increase their draft. At low tide, when the laden craft are riding very low in the water, the rescue crew pulls the cables taut. The lift craft release their ballast with the rising tide, thereby raising the sunken ship the combined height of the new tide and the decreased draft of the lift boats. The downed ship, dangling as if in an underwater hammock, is then regrounded in shallower water, where the whole process is begun anew, and subsequently repeated until the sunken ship rests above water and can be drained.

By employing one or a combination of ship-raising techniques, salvagers have the technology to recover virtually any vessel. Often, however, physical and financial risks conspire to make a raising unfeasible, which is why even the world's most famous shipwreck, the *Titanic,* will probably never be raised.

?

How do they know how long a second is?

THE OLD WAY (AND the way most of us still use) is derived from the time it takes for a complete day to pass. For centuries we have split a day into twenty-four hours, an hour into sixty minutes and a minute into sixty seconds. But though the Earth's movements from day to day are constant enough to schedule business lunches by, they do vary minutely. Military defense systems, communications satellites and space travel require literally split-second timing. Without it, for example, a few instants of the Academy Awards would be lost in space instead of beaming from Hollywood to the network satellite, to the network affiliate, and to you. The ramifications in defense and space travel are significantly more sobering.

In the late 1940s, Isidor Rabi, a physicist at Columbia University, found that certain atoms vibrate consistently at exactly the same rates. He suggested that atoms could make good clocks. Electronically counting the tiny vibrations of millions of atoms in fourteen-foot vacuum tubes, U.S. and British scientists found that the atom ^{133}cesium (the isotope of cesium with the atomic weight 133) was one of the most consistent and easy to measure. After hundreds of tests carried out over several years, they determined that it vibrates exactly 9,192,631,770 times in what they considered an average second.

The General Conference of Weights and Measures, an international treaty organization, met in Paris in 1967 and officially changed its definition of a second. No longer measured by the Earth's movements, the second became the exact length of time occupied by 9,192,631,770 vibrations of a ^{133}cesium atom. The beam of vibrating cesium atoms was essentially the pendulum of an incredibly accurate clock.

How accurate? The best atomic clocks in the world will be off by one second after 300,000 years of steady running, compared with a quartz wristwatch, which will (theoretically, of course) miss the mark by at least a million seconds, or 278 hours, over the same time period.

?

How do they decide what to charge for a minute of advertising during the broadcast of the Super Bowl?

IN ANY GIVEN YEAR, advertising rates for the Super Bowl are set by the network that owns the broadcast rights for that year's game. There are no strict equations for setting a price. Instead, the network will decide what price it thinks the market can bear and charge accordingly. In 1991, ABC figured $800,000 was about right for a thirty-second spot, the standard commercial time unit. ABC sold all fifty-six of its thirty-second advertising slots that year, for a grand total of twenty-eight minutes of network commercials. At $1.6 million per minute, that translated into $44.8 million in advertising revenues for the network. Not bad for an hour of football.

Although the networks have no absolute formula for determining the price of commercial time, they do follow one rough guideline: the price never goes down. The only time the cost of Super Bowl advertising declined was in 1977, during a recession, when the price fell by $25,000 from the previous year, from about $150,000 for thirty seconds to roughly $125,000. In the next two years the price jumped up to just over $200,000. Rates have risen every year since. The increase might be as much as 15 percent or as little as 5 percent from one year to the next, but there is always an increase.

Inflation accounts only partially for the skyrocketing price of Super Bowl ad time. The real reason commercial slots have become so costly is that the Super Bowl has developed into a showcase event for corporations. The big manufacturers in every industry, from computers to beer to athletic footwear, have seized upon it as the time to unveil new products and launch new ad campaigns. Some companies now spend as much as 30 percent of their annual advertising budget on the Super Bowl. The single most compelling reason for this corporate strategy is the game's huge television viewership. It almost always draws the largest audience of any show broadcast during the year. In 1991, more than 100 million people in the United States, and hundreds of millions more abroad, tuned in. According to Nielsen Media Research, Super Bowls account for five of the ten all-time most popular programs in American television history. Advertisers know there will be a large audience for their spots. As one executive whose company has spent a bundle in recent years explained it, "The game is not a typical football event that caters to male viewers. It really is the convening of American men, women, and children, who gather around the sets to participate in an annual ritual."

Despite all the hoopla surrounding the game, the size of the viewing audience has actually been shrinking in recent years. The highest-rated Super Bowl of all time was the 1982 game, which pulled down a 49.1 rating. (Each rating point represents 1 percent of all U.S. households with televisions.) By 1990, the rating had slipped to 39. But that still represented 35.9 million households and more than 100 million individuals. Considering the size of the audience, Super Bowl advertising is not really so expensive. During the height of its popularity, for instance, *The Cosby Show* commanded $300,000 for a thirty-second ad spot, and the audience was less than half that of the Super Bowl. On the other hand, while the peak *Cosby* episodes were usually pretty good television shows, the Super Bowl often turns out to be a dud of a football game.

?

How do they know what a particular dinosaur ate?

WE GROW UP ABSORBING explicit illustrations from picture books of sauropods nibbling leaves and tyrannosaurs ripping flesh, and yet some paleontologists maintain that these images are largely conjecture. We can speculate about what the dinosaurs ate, but in fact no one knows for sure, just as we can only imagine what color the dinosaurs were and what sounds they made. Our guesswork about what dinosaurs consumed is based on parallels with mammals we know today, yet dinosaurs were of course extremely different from any creature we have ever known. Bearing this in mind, we can look at some different species and simply theorize about their diet, largely by peering into their mouths at the size and shape of their teeth.

Take *Diplodocus*, for instance, which had stubby, pencil-like teeth jutting slightly forward from the jaw and located mostly at the front of the mouth. It would seem impossible for this slim-limbed creature to have chomped down on an armored dinosaur or even gnawed down a tree. Instead, it probably used its teeth like a sieve to obtain tiny organisms from fresh water and, since excavated teeth of this species show signs of wear, it may also have combed or cropped twigs and leaves.

Tracks show that the huge, peaceful, herbivorous sauropods ("lizard feet") like *Diplodocus*, *Apatosaurus*, and *Brachiosaurus* may have traveled in herds. With their long necks they could have grazed easily among the treetops, some forty feet high. They would have spent most of their waking hours munching, because an eighty-ton brachiosaur, if cold-blooded, would have required some three hundred pounds of vegetation a day to sustain itself, if warm-blooded it would have had to consume a ton of greens each day. Since its teeth were little help in breaking down foodstuff, one theory goes that the sauropods swallowed rocks, which ground food to a pulp inside a gizzard. Thousands of such stones, known as gastroliths, have been found along with the bones of these reptiles.

While the sauropods may have held sway over the forest canopy, other dinosaurs were equipped to feed on the lower tiers of woods and in swamps. *Iguanodon* had numerous sharp cheek teeth for consuming twigs and leaves. It roamed warm swampy areas and probably ate tough marsh plants such as horsetails. Horned dinosaurs such as *Triceratops* had a strong beak for tearing down and stripping tough, woody vegetation. Its many sharp teeth meshed like scissors to masticate spiky palm fronds, magnolias, and fibrous plants. A long, thick muscle that stretched from the lower jaw to the frill at the back of the neck added strength to its jaws.

A very different creature, whose teeth are a notable indicator of life-style, is the well-known *Tyrannosaurus rex*. This animal's stupendous teeth measured seven and a quarter inches in length and, like steak knives, were perfectly designed to slice flesh. They were narrow and serrated, curving slightly inward to clench hold of prey. Some scientists believe that this enormous dinosaur, which weighed about seven tons, was one of the most ferocious hunters that ever lived, bringing down even tough horned and armored dinosaurs. Others maintain that *Tyrannosaurus* was too large and clumsy to tackle live prey and that it merely lumbered about in search of carrion.

Other carnosaurs, or "flesh lizards," such as the allosaurids and megalosaurids, also had saw-edged fangs, powerful jaws, and two- or three-fingered hands which could have been used for grasping prey. Probably the earlier species, which were smaller and more active than the later ones, were nimble enough to run down a browsing plant-eater.

There has been just one instance in which paleontologists uncovered the actual stomach contents of a dinosaur. In 1922 an *Anatosaurus* was found with a belly full of twigs and pine needles. At the time the finding was ignored because the scientific community believed that anatosaurs and other duck-billed dinosaurs ate like ducks, that is, aquatic plants. Over four decades later John H. Ostrom of Yale University reintroduced the find and demonstrated that the anatosaur, and other hadrosaurs, were not aquatic. It so happens that the teeth of the anatosaur would have been well suited for chewing its dinner. Behind the bill of a hadrosaur lay rows of packed teeth, and beneath them, new teeth growing up to replace those that wore out. Hadrosaurs munched on hard dry-land fare, such as branches and seeds, and over the course of a lifetime would have grown and put to work some two thousand teeth.

One hopes that paleontologists will be fortunate enough to find other species of dinosaurs mummified with a full stomach, but until they do, much of what we suppose about these unique creatures must hark from the land of make-believe.

?

How do they measure the heat of distant stars?

WHEN YOU LOOK at the grill on an electric stove as it heats up, you see it turn from black to bright red, and you know by the color that it's hot enough to blister your fingers. Astronomers determine stars' temperatures in much the same way—by studying their colors. Of course, stars are a lot hotter than a stove: the coolest are red. Progressively hotter ones are orange, yellow, green, blue and, finally, violet. Blue and violet stars are rare; red stars are common because, for one thing, they burn out more slowly.

The next time you are out on a clear night (a dry winter sky far from city lights is best) take a look at Orion, a large constellation suggesting the shape of a hunter brandishing a club and a lion skin. The bright red star at Orion's shoulder is Betelgeuse (pronounced "beetle juice"); the foot star of Orion is Rigel, a clear blue star. These colors are visible with the naked eye.

Astronomers determine the color of light emitted by a star with equipment called a spectrophotometer. The light waves are then characterized by wavelength. Astronomers use a mathematical formula that involves the speed of light and some other universal constants to convert wavelengths to temperature: the longer the wavelength, the lower the energy, the cooler the star. With that formula, they have calculated that Betelgeuse is about five thousand degrees Fahrenheit, and Rigel is about four times as hot. Knowing how hot a star is, astronomers can begin to understand its mass, its distance from Earth and how much it has aged.

?

How do they know there's going to be another ice age in two thousand years?

NO ONE KNOWS for certain when the next ice age will begin, but geological evidence indicates that it could start within two thousand years. During the past 2 million years, ice ages have been coming and going according to a fairly regular schedule. They seem to last about 100,000 years or so, and are usually separated by short interglacial periods lasting about eight thousand to twelve thousand years. We are living in an interglacial period that began when the last ice age ended, roughly ten thousand years ago. The historical model would therefore suggest that the next ice age is probably already on its way, and should arrive in a couple thousand years.

The term *ice age* has come to refer to the peaking of the ice sheets during the Pleistocene Epoch. The Pleistocene epoch began about 1.5 million years ago; it was a time of regularly reappearing ice ages. The end of the most recent ice age, about ten thousand years ago, marked the close of the Pleistocene and the start of the modern, Holocene Epoch, the interglacial epoch during which the human race has flourished. At the height of the last ice age the average global temperature was about five degrees Celsius cooler than it is today. One third of the Earth's surface, including northern Europe, Canada, Greenland, Antarctica, parts of the United States, and areas of Australia, New Zealand, and South America, was blanketed in ice sheets up to ten thousand feet thick. The huge glaciers contained 5 percent of the world's water, which caused sea levels to be lowered by as much as four hundred feet.

The rising and falling of sea levels that comes with global glaciation has helped scientists chart the succession of ice ages. Coral grows only in warm waters close to the surface of the ocean. When the sea level dropped with the last ice age, coral dropped with it. As the glaciers melted and the sea began to rise again, new coral grew on top of the old. Each level of coral growth marks what was once the surface of the sea. By dating coral growth scientists have been able to put together a fairly detailed chronology of the last ice age.

Another way scientists have been able to date ice ages is by examining the fossil record. The presence of leafy plants, for example, indicates a warm climate. By studying and dating vegetable fossils, scientists can tell what the global climate was like during a certain time. One thing they have learned is that during the interglacial period prior to the last ice age, the Earth was warmer than it is today. The strongest evidence for this belief comes from rock sediments from the ocean floor. The levels of carbon dioxide in such sediments indicate that there was more carbon dioxide in the atmosphere during the previous interglacial period than there currently is. High atmospheric carbon dioxide levels create a greenhouse effect, whereby radiation from the sun is trapped inside the Earth's atmosphere, causing the global temperature to rise. During the last century, the greenhouse effect has caused global warming at an annual rate of 0.3 to 0.4 degrees Celsius. Sometime during the next century the global temperature should reach as high as it was during the last interglacial.

When global temperature rises, the polar ice caps melt. If the ice caps were to thaw, they would surge toward the ocean. As a result, massive icebergs would be discharged into the sea. The flow of ice into the sea would raise sea levels around the globe, which would cause total inundation of coastal areas, which in turn would cause more ice to spill into the water. As all this ice entered the ocean, ocean temperatures would drop, causing more sea ice to form. Since ice reflects rather than absorbs heat, the extra sea ice would radiate heat back out of the atmosphere, causing temperatures to drop further. More ice would then form. And so on. Once the ice gets going it pretty much takes care of itself. In this way, some scientists speculate, one unlikely and unwholesome effect of global warming will be to bring about the next ice age.

?

Sources

How do they detect counterfeit bills?

Bureau of Engraving and Printing, (202) 447-0193.
Crane and Company, (413) 684-2600.
"Know Your Money," distributed by the Department of the Treasury and the United States Secret Service.

How do they keep you from registering to vote in more than one jurisdiction?

Kimberly, Bill. National Clearinghouse, Federation Election Commission, (202) 376-3155.

How do they get ships through the Panama Canal?

Encyclopedia Americana. Danbury, Conn.: Grolier Inc., 1988.
McCullough, David, author of *The Path Between the Seas: The Creation of the Panama Canal, 1870–1914*. New York: Simon and Schuster, 1977.

1987 Information Please Almanac®, Atlas and Yearbook, The, 40th ed. Boston: Houghton Mifflin.
World Book Encyclopedia. Chicago: World Book, Inc., 1985.

How do they make a breed of dog "official" at the American Kennel Club?

Derr, Mark. "The Politics of Dogs." *The Atlantic*, March 1990.

How do they write headlines at the *New York Post*?

Colasuonno, Lou. *New York Post*, 210 South Street, New York, N.Y. 10002; (212) 815-8500.

How do they find arbitrators for baseball contract arbitrations?

Brown, Laura. American Arbitration Association, (212) 484-4129.
Players Association, (212) 826-0808.
Schack, Arthur, attorney. Office of the Commissioner, (212) 339-7800.

How do they make margarine taste like butter (almost, anyway)?

Bradley, Dr. Robert L., Jr., professor of food science. University of Wisconsin, Madison.
Freund, Dr. Peter, manager of product applications, and Pamela J. Kokot, product manager. Both at Chr. Hansen's Laboratory, Inc., Milwaukee, Wis.

How do so many Japanese play golf in a country with so few golf courses?

"Personal Affairs." *Forbes*, July 11, 1988.
Reilly, Rick. "Japanese Golf." *Sport Illustrated*, August 21, 1989.

How do they set the price on a new public stock offering?

Downes, John, and Jordan Goodman. *Dictionary of Finance and Investment Terms*. Hauppauge, N.Y.: Barron, 1985.
"The Billionaires." *Fortune*, September 9, 1991.

How do they know Jimmy Hoffa is dead?

"Hoffa Outgunned." *Time*, July 5, 1982.
"Jimmy Hoffa Is Legally Dead." *Newsweek*, August 9, 1982.
New Encyclopaedia Britannica. Chicago: Encyclopaedia Britannica, 1986.
Sifakis, Carl. *The Encyclopedia of American Crime*. New York: Facts on File, 1982.
Treaster, Joseph B. "Hoffa Ruled 'Presumed' Dead." *The New York Times*, December 10, 1982.
Witkin, Gordon. "Jimmy Hoffa's Deadly Lunch." *U.S. News and World Report*, July 11, 1988.

How do they get rid of gallstones without operating?

Good Housekeeping, September 1989.
National Center for Health Statistics, (301) 436-8500.
Saturday Evening Post, The, May-June 1981, October 1988, May-June 1989.
Yin, Dr. Lillian. Food and Drug Administration, (301) 427-1180.

How do they make telephones for the deaf?

Dreyfus, Barbara. Ultratec, Madison, Wis.; (800) 233-9130.
National Association for the Deaf, (301) 587-1788.
National Information Center on Deafness, (202) 651-5051.

How do banks make money off of credit card purchases even when you don't pay interest?

Anglin, David. Visa U.S.A., Research Department, P.O. Box 8999, San Francisco, Calif. 94128; (415) 513-1103.

"Heart from Hart." *United States Banker.* Kalo Communications, Inc., August 1991, national edition.

Wasserman, Gail, director public affairs. American Express Travel Related Services Company, Inc., American Express Tower, World Financial Center, New York, N.Y. 10285-3130; (212) 640-2675.

How do they choose cartoons for *The New Yorker*?

Lorenz, Lee, art editor. *The New Yorker*, New York, N.Y.

Menand, Louis. "A Friend Writes." *The New Republic*, February 26, 1990.

How do they make rechargeable batteries?

Kluzik, John D. Lighting Information Center, General Electric Company, Cleveland, Ohio 44112; (216) 266-8346.

Mantell, Charles L. *Batteries and Energy Systems.* New York: McGraw-Hill, 1983.

How do they set the minimum bid on a painting that is up for auction?

Hughes, Robert. "Sold." *Time*, November 27, 1989.

Sotheby's, Old Master Paintings and Drawings Department, New York, N.Y. (212) 606-7230.

How do antilock brakes work?

Dinan, Jack. General Motors, Dearborn, Mich.; (313) 986-5717.

How do they know television viewers don't cheat during the Nielsen ratings?

Buono, Diane. Nielsen Media Research, 1290 Avenue of the Americas, New York, N.Y. 10019; (212) 708-7500.

How do they manufacture genes?

"Amazing Gene Machine, The," *U.S. News and World Report*, July 16, 1990.

Benson, Don, senior sales specialist. Ependorf North America, (908) 526-1389.

"Gene Therapy." *Mirabella*, December 1991.

"Genetic Age, The," *Business Week*, May 28, 1990.

"Genetic Engineering." *Encyclopedia of Chemical Technology*, 3rd ed. New York: Interscience Publishers.

New Encyclopaedia Britannica. Chicago: Encyclopaedia Britannica, 1986.

How do they stop you from getting a credit card under a false name?

Crone, Kenneth R., vice president. Visa U.S.A., San Francisco, Calif.

Ferrell, Jack, sales manager. Fraud Management Division, First Data Resources, Omaha, Neb.

Mandell, Lew, author of *The Credit Card Industry: A History*. Boston: Twayne Publishers, a division of G. K. Hall, 1990.

"Visa Risk Management and Security," a booklet from Visa U.S.A.

How do members of Congress make themselves sound articulate in the *Congressional Record*?

Nicestrom, Dwayne, editor. *Congressional Record*, Washington, D.C.; (202) 224-6810.

How do they "make" veal?

Ellis, Merle. *Cutting Up in the Kitchen.* San Francisco: Chronicle Books, 1976.

Evans, Travers Moncure, and David Greene. *The Meat Book.* New York: Charles Scribner's Sons, 1973.

National Livestock and Meat Board, Chicago, Ill.; (312) 467-5520.

How do they determine the profit of a Hollywood movie?

Cohn, Lawrence. *Variety,* New York, N.Y.; (212) 779-1100.

Creer, John. Exhibitor Relations, Los Angeles, Calif.; (213) 657-2006.

Hanover, Debbie. Entertainment Data, Inc., Los Angeles, Calif.; (213) 658-8300.

National Amusements, Inc., Dedham, Mass., (617) 461-1600.

How do they decide who gets nominated for Oscars?

Brown, Peter N., and Jim Pinkerton. *Oscar Dearest.* New York: Harper and Row, 1987.

Whirton, Bob. Academy of Motion Picture Arts and Sciences, (213) 385-5271.

How do they make nonalcoholic beer?

Brewer's Guardian, March 1987.

Modern Brewery Age, May 16, 1988.

Sharalambas, George. Brewing Technical Services, Anheuser-Busch, Inc., (314) 577-3222.

How do they know there was an ice age?

"Learning the Language of Climate Change." *Mosaic* 19, Fall/Winter 1988.

Ray, Louis L. *The Great Ice Age*. Washington, D.C.: U.S. Department of the Interior/U.S. Geological Survey.

How do they correct nearsightedness in ten minutes?

"Goodby Glasses?" *Consumer Reports*, January 1988.
Hacinli, Cynthia. "Eyes Can See Clearly Now: Surgery for the Nearsighted." *Mademoiselle*, June 1990.
Krantz, Paul. "Don't Toss Your Glasses—Yet." *Better Homes and Gardens*, June 1990.
Lynn, Michael, director. The Statistical Coordinating Center for the Prospective Evaluation of Radial Keratotomy, Emory University School of Public Health, Atlanta, Ga.

How do lawyers research members of a jury before deciding whether to challenge them?

Colson, Bill, Lisa Blue, and Jane N. Saginaw. *Jury Selection: Strategy and Science*. Callaghan Trial Practice Series. Deerfield, Ill.: Callaghan, 1986.
Duboff, Robert. Decision Research Corporation, 33 Hayden Avenue, Lexington, Mass. 02173; (617) 861-7350.
Owen, Robert. Owen and Davis, 605 Third Avenue, New York, N.Y. 10158; (212) 983-6900.
Sanders, William H. "The Voir Dire in a Civil Case." *University of Missouri Kansas City Law Review*, Vol. 57, No. 4. Kansas City: University of Missouri Kansas City School of Law, 1989.

How does an electric eel generate electricity?

Bridges, William. *The New York Aquarium Book of the Water World*. New York: American Heritage Press, New York Zoological Society, 1970.
Burton, Maurice, and Robert Burton. *Encyclopedia of Fish*. New York: American Museum of Natural History, BPC Publishing Ltd., 1968.

Dozier, Thomas A. *Dangerous Sea Creatures*. Based on the television series *Wild, Wild World of Animals*. Time-Life Films, Inc., 1974.

Ferraris, Dr. Carl. Department of Ichthyology, American Museum of Natural History, 79th Street and Central Park West, New York, N.Y. 10024.

Gray, Peter, ed. *Encyclopedia of Biological Sciences*. New York: Reinhold Publishing Corp., 1961.

Marshall, N. B. *The Life of Fishes*. New York: Universe Books.

Schultz, Leonard P. *The Ways of Fishes*. New York: D. Van Nostrand Co., Inc.

How does an invisible fence keep your dog from straying?

Invisible Fencing, 724 West Lancaster Avenue, Wayne, Pa. 19087; (215) 732-4100.

How does a thief break into your car and drive off with it in less than a minute?

FBI Auto Theft Task Force, 26 Federal Plaza, New York, N.Y. 10278; (212) 335-2700.

How do they get the oat bran out of oats?

Cornell Medical College, 1300 York Avenue, New York, N.Y. 10021; (212) 746-5454.

Encyclopaedia Britannica. Chicago: Encyclopaedia Britannica, 1991.

General Mills, 1 General Mills Blvd., Minneapolis, Minn. 55426; (612) 540-2311.

Institute of Human Nutrition, Columbia University, 701 West 168th Street, New York, N.Y. 10032; (212) 305-6991.

How do they get the bubbles into seltzer?

Encyclopaedia Britannica. Chicago: Encyclopaedia Britannica, 1991.

Mid-Atlantic Cans Association, (413) 586-8450.

Ostendarp, Beth, communications and public affairs specialist. Cadbury-Schweppes, (203) 329-0911.

How do they teach guide dogs to cross at the green light?

Siddall, Elane. The Guide Dog Foundation, Smithtown, N.Y.; (516) 265-2121.

How do they measure dream activity?

Cartwright, Rosalind Dymond. *A Primer on Sleep and Dreaming.* New York: Addison-Wesley, 1978.

Feldman, Professor Sam. Department of Neurophysiology, New York University, (212) 998-7852.

Hobson, Professor Allan. Harvard University, (617) 734-1300.

LaBerge, Stephen. *Lucid Dreaming.* New York: St. Martin's Press, 1985.

Luce, Gay Caer, and Julius Segal. *Sleep and Dreams.* New York: Coward-McCann, 1966.

Rappaport, Dr. Joyce. Sleep Research Laboratory, New York University Medical School, (212) 340-6407.

How do they decide whose obituary gets published in *The New York Times*?

The New York Times, (212) 556-1234.

Thompson, Marty, managing editor. Associated Press, (212) 621-1610.

How do they know there are more stars in the sky than grains of sand on an ocean beach?

"Sand and Stars." *The New York Times*, Science Times section, March 19, 1991.

Tyson, Dr. Neil D. Department of Astrophysical Sciences, Princeton University, Princeton, N.J. 08544; (609) 258-2303.

Tyson, Neil D. "A Sentimental Journey to the Googolplex." *Star-Date*, February 1982.

How do they decide where to put a new book in the Library of Congress?

Library of Congress, Public Affairs Office, (202) 707-2905.

How do they discover a new drug?

Werth, Barry. "Quest for the Perfect Drug." *New England Monthly*, February 1990.

How do they get honey out of a honeycomb?

Encyclopedia Americana. Danbury, Conn.: Grolier Inc., 1988.

Mace, Herbert. *The Complete Handbook of Bee-Keeping*. New York: Van Nostrand Reinhold, 1976.

Morse, Roger A., and his book *The Complete Guide to Beekeeping*, rev. ed. New York: E. P. Dutton, 1974.

How do they know what caste a resident of Bombay belongs to?

Jacob, Louis. Asian Division, South Asian Section, Library of Congress, (202) 707-5600.

Kolenda, Pauline. *Caste in Contemporary India*. Prospect Heights, Ill.: Waveland Press, 1985.

Olson, Mancur. *The Rise and Decline of Nations*. New Haven, Conn.: Yale University Press, 1982.

How do they dry-clean clothes without getting them wet?

Grimes, William. "Every Day, a Magic Show at the Dry Cleaners." *The New York Times*, November 14, 1991.

Ray, C. Claiborne. "Dry Cleaning." *The New York Times*, November 5, 1991.

Spotting: The Art of Stain Removal. Neighborhood Cleaners Association, New York School of Drycleaning, 116 East 27th Street, New York, N.Y. 10016-8998; (212) 684-0945.

How do they make glow-in-the-dark toys?

Brown, Harvey E. *Zinc Oxide: Properties and Applications.* New York: International Lead Zinc Research Organization, 1976.

Encyclopedia Americana. Danbury, Conn.: Grolier Inc., 1988.

Majka, Robert. Canrad-Hanovia, Inc., 100 Chestnut Street, Newark, N.J. 07105; (201) 589-4300.

Ruddick, Douglas H. Zinc Corporation of America, 300 Frankfort Road, Monaca, Penn. 15061-2295; (412) 774-1020.

Wagner, Dr. Peter J. Michigan State University, (517) 355-1855.

How do they predict when a fine wine will be ready to drink?

Adams, Leon D. *The Commonsense Book of Wine.* New York: McGraw-Hill, 1986.

Johnson, Hugh. *Hugh Johnson's Encyclopedia of Wine.* New York: Simon and Schuster, 1983.

Kressman, Edward. *The Wonders of Wine.* New York: Hastings House, 1968.

Tchelistcheff, André, vintner, (707) 224-5502.

How do they know how much an aircraft carrier weighs?

Austin, William H., III. Department of the Navy, Naval Sea Systems Command, Washington, D.C. 20362-5101; (703) 602-9083.

How do they decide what or who gets honored on U.S. postage stamps?

Arroyo, Norma. U.S. Postal Service, (202) 268-2323.

How do emperor penguins stay warm in Antarctica?

Burton, Robert. *Bird Behavior*. New York: Alfred A. Knopf, 1985.

Pearce, Q. L. *Tidal Waves and Other Ocean Wonders*. Englewood Cliffs, N.J.: Julian Messner, 1989.

Simpson, George Gaylord. *Penguins: Past and Present, Here and There*. New Haven, Conn.: Yale University Press, 1976.

Stonehouse, Bernard. *Penguins*. New York: Golden Press, 1968.

How do they treat drug addiction with acupuncture?

Bullock, Milton L., Patricia D. Culliton, and Robert T. Ollander. "Controlled Trial of Acupuncture for Severe Recidivist Alcoholism." *The Lancet*, June 24, 1989.

"How Does Acupuncture Work?" *British Medical Journal*, Vol. 283, September 19, 1981.

O'Rourke, Ellen, licensed acupuncturist, (413) 256-8320.

Requena, Yves. "Acupuncture's Challenge to Western Medicine." *Advances*, Vol. 3, No. 2, Spring 1986.

How do tightrope walkers stay on the wire?

Quiros, Angel, high-wire artist for Ringling Brothers; c/o Beth Painter, Public Relations for Ringling Brothers, (703) 448-4118.

How do they recycle newspapers?

Cogoli, John. *Photo Offset Fundamentals*. Bloomington, Ill.: McKnight, 1973.

Commoner, Barry. *Making Peace with the Planet.* New York: Pantheon Books, 1990.
"Recycled Paper Facts and Figures," a fact sheet of the National Recycling Coalition, Inc., Washington, D.C.
Steger, Will, and Jon Bowermaster. *Saving the Earth: A Citizen's Guide to Environmental Action.* New York: Alfred A. Knopf, 1990.
"Tough Business of Recycling Newsprint, The," *The New York Times*, January 6, 1991.

How does a magician pull a rabbit out of a hat?

Christopher, Milbourne. *The Illustrated History of Magic.* New York: Thomas Y. Crowell, 1973.
Dawes, Edwin A., and Arthur Setterington. *The Encyclopedia of Magic.* New York: W. H. Smith Publishers, 1986.

How do they make long-life milk?

Dairymen/Farm Best Division, 11 Artley Road, Savannah, Ga. 31408; (912) 748-6185.

How do they pick Pulitzer Prize winners?

Bates, J. Douglas. *The Pulitzer Prize.* New York: Birch Lane Press, 1991.
Hohenbert, John. *The Pulitzer Prizes.* New York: Columbia University Press, 1974.
The New York Times, May 12, 1991, and January 24, 1988.

How do they make crack?

Brown, Ida, public affairs officer. Drug Enforcement Administration, Washington, D.C., (202) 307-1000, ext. 4.
Morales, Edmundo. *Cocaine: White Gold Rush.* Tucson, Ariz.: University of Arizona Press, 1989.

How do they measure the unemployment rate?

Cohany, Sharon R., economist. U.S. Department of Labor, Bureau of Labor Statistics, (202) 523-1944.

How the Government Measures Unemployment, Report 742, U.S. Department of Labor, 1987.

How do they interview and hire simultaneous interpreters at the United Nations?

Aimé, Monique Corvington, chief of Interpretation Service, United Nations, (212) 963-8233.

UN Chronicle, September 1991.

Visson, Lynn. *From Russian into English: An Introduction to Simultaneous Interpretation*. Ann Arbor, Mich.: Ardis Publishers, 1991.

How do they spread a computer virus?

Fites, Philip, Peter Johnston, and Martin Kratz. *The Computer Virus Crisis*. New York: Van Nostrand Reinhold, 1989.

McAfee, John, chairman of the Computer Virus Industry Association, and Colin Haynes. *Computer Viruses, Worms, Data Diddlers, Killer Programs, and Other Threats to Your System: What They Are, How They Work, and How to Defend Your PC, Mac, or Mainframe*. New York: St. Martin's Press, 1989.

Schram, Martha. McAfee Associates in Santa Clara, Calif.

How do they get the caffeine out of coffee?

Heiss, Bob, co-owner of the Coffee Gallery in Northampton, Mass.

"Key Facts About Decaffeination," a brochure from the Coffee Development Group in Washington, D.C.

Kummer, Corby. "Is Coffee Harmful?" *Atlantic*, July 1990.

Lewis, Charles, a coffee trader for Cofinco, New York, N.Y.

National Coffee Association, New York, N.Y.

How do they test golf balls before offering them for sale?

Forbes, William T., communications manager. United States Golf Association, (908) 234-2300.

Spaulding Corporation, Chicopee, Mass., (413) 536-1200.

How do they see a black hole?

Wald, Robert M., professor at the University of Chicago and author of *Space, Time and Gravity.* Chicago: University of Chicago Press, 1977.

How do they get the music onto a CD?

Birchall, Steve. "The Magic of CD Manufacturing." *Stereo Review*, October 1986.

Booth, Stephen A. "The Digital Revolution." *Rolling Stone*, January 20, 1983.

Dumaine, Brian. "The Compact Disk's Drive to Become the King of Audio." *Fortune*, July 8, 1985.

Greenleaf, Christopher. "CD Boom! Compact Disc Pressing Plants Are Spreading Across North America." *Stereo Review*, June 1987.

Seligman, Gerald. "Saved! How Classic Rock Tracks Are Kept Forever Young on CD." *Rolling Stone*, September 11, 1986.

Strawn, John, series editor. *The Computer Music and Digital Audio Series.* Vol. 5: *The Compact Disc: A Handbook of Theory and Use.* Madison, Wis.: A-R Editions, Inc., 1989.

How do they make the Hudson River drinkable?

Fairbanks, Douglas, chief operator. Poughkeepsie Water Treatment Plant, (914) 451-4173.

Lee, Cara. Scenic Hudson, (914) 473-4440.

Shaw, Susan, environmental engineer. Public Water Supply Section, Environmental Protection Agency, (212) 264-4448.

How do they pack Neapolitan ice cream?

Belanger, Michael. Merriam-Webster, Inc., Springfield, Mass., (413) 734-3134.

Encyclopedia Americana. Danbury, Conn., Grolier Inc., 1989.

Erickson, Don. Public Affairs, H. P. Hood Company, Boston, Mass.; (617) 242-0600.

How do they make sure condoms won't break?

Beheler, Linda. Bloom Public Relations, Inc., New York, N.Y.; (212) 370-1363, ext. 403.

Men's Fitness, May 1991.

How do they make stainproof carpets?

E. I. du Pont de Nemours and Company, Wilmington, Del.; (800) 438-7668.

How do they teach a bear to ride a bicycle?

Ringling Brothers and Barnum & Bailey Circus.

How do they measure the ozone layer?

Lindley, D. "Ozone Hole Heading South Prematurely." *Nature*, November 9, 1989.

McGraw-Hill Encyclopedia of Science and Technology. New York: McGraw-Hill, 1992.

Monastersky, R. "Depleted Ring Around Ozone Hole." *Science News*, November 18, 1989.

Prather, M. J., and R. T. Watson. "Stratospheric Ozone Depletion and Future Levels of Atmospheric Chlorine and Bromine." *Nature*, April 19, 1990.

Schoeber, Mark R. "Arctic Ozone Succumbs to Chemical Assault." *Science News*, March 24, 1990.

How do they count calories in food?

Katch, Frank I., and William D. McArdle. *Nutrition, Weight Control, and Exercise*, 2d ed. Philadelphia: Lea and Febiger, 1983.

How do they know how much money will come out of a cash machine?

Jensen, Roger, senior product manager. NCR Corporation, Dayton, Ohio.

How does the Pennsylvania Department of Agriculture get mentioned on cereal boxes?

Readel, Dick, and Dyan Yingst. Pennsylvania Department of Agriculture, Commonwealth of Pennsylvania, (717) 787-5085.

United States Department of Agriculture, (202) 447-2791.

How do they detect wind shear?

Aircraft Owners and Pilots Association, Frederick, Md.; (301) 695-2000.

Federal Aviation Administration, Washington, D.C.; (202) 267-8521.

McCarthy, John. National Center for Atmospheric Research, Boulder, Colo.; (303) 497-8822.

How do they make fake photos on magazine covers?

Guiseffi, Ralph. AGT Printing, Washington, D.C.; (202) 955-2491.

Murphy, Eric. Tru-Color, Greenfield, Mass.; (413) 774-7901.

How do they make shredded wheat?

Nabisco Brands, Inc., East Hanover, N.J. 07936-1928.

How do they write words across the sky in airplane exhaust?

Arkin, Mort. Skytyping Inc., Marine Air Terminal, Flushing, N.Y.

How do they determine who gets into Harvard?

Colodny, Mark M. "Building Observatories Helps: Writing Backwards Doesn't." *U.S. News and World Report*, August 25, 1986.

Fischgrund, Tom, ed. *Barron's Top 50*. Hauppauge, N.Y.: Barron, 1991.

Greene, Howard. *Scaling the Ivy Wall.* Boston: Little, Brown, 1987.

Larew, John. "Why Are Droves of Unqualified, Unprepared Kids Getting into Our Top Colleges?" *The Washington Monthly*, June 1991.

How do chameleons change color?

Bellairs, Angus. *The Life of Reptiles.* New York: Universe Books, 1970.

Martin, James. "The Engaging Habits of Chameleons Suggest Mirth More Than Menace." *Smithsonian*, June 1990.

How do they make high-definition television?

Boston Globe, The, March 18, 1991.

Business Week, December 21, 1987.

HDTV 1125/60 Group, The, 1615 L Street, Washington, D.C. 20036; (202) 659-1992.

Newsweek, April 4, 1988.

Popular Mechanics, July 1986.

Popular Science, January 1987.

Time, December 21, 1987.

U.S. News and World Report, January 23, 1989.

How do they wrap Hershey's Kisses?

Fortna, Bob, Hershey plant manager. Hershey Foods, Hershey, Penn.

Great Chocolate Story, The, Hershey Foods promotional videotape.

Hershey Foods, Office of Public Information, Hershey, Penn.

How do they teach a computer to recognize your voice?

IBM, 1 Old Orchard Road, Armonk, N.Y. 10504; (914) 765-1900.

How do they know who has won an election before the polls close?

Harper's, July 1984.

New Republic, The, March 28, 1988, and November 28, 1988.

Newsweek, March 19, 1984.

U.S. News and World Report, November 12, 1984.

How do they train structural steelworkers to walk on unprotected beams five hundred feet in the air?

Kelly, John, coordinator of the Joint Apprenticeship and Trainee Committee of Ironworkers Locals 40 and 361, New York, N.Y.

Kittle, Martin, business manager, Local 806, Structural Steel and Bridge Painters, New York, N.Y.

How do they find heart or liver donors?

Benenson, Esther L., publications director. United Network for Organ Sharing, Richmond, Va., (804) 330-8500.

Business Week, August 28, 1989.

Newsweek, December 11, 1989.

How do they get the cork into a bottle of champagne?

Kaufman, William. *Champagne.* New York: Viking Press, 1973.
Larousse Wines and Vineyards of France. New York: Arcade Publishing, Inc., 1990.
Lichine, Alexis. *New Encyclopedia of Wines and Spirits.* New York: Alfred A. Knopf, 1982.

How do they make Hostess Twinkies stay fresh for years and years?

Farrell, Patrick. Continental Baking Company, St. Louis, Mo.; (314) 982-2261.

How does the SEC know when someone's doing inside stock trading?

Business Week, May 26 and December 1, 1986.
Newsweek, September 19, 1988.
Securities and Exchange Commission, Public Affairs, (202) 272-2650.
Time, July 14 and December 1, 1986.
U.S. News and World Report, December 22, 1986, and September 26, 1988.

How do they decide what goes on the cover of *People* mgazine?

Toepfer, Susan, assistant managing editor. *People*, New York, N.Y.

How do they put the smell into scratch-and-sniff advertising?

Conklin, Jim. Lawson Mardon, Inc.

How does sunscreen screen out the sun?

Agin, Patricia. Schering Plough HealthCare Products, Memphis, Tenn. 38151; (901) 320-5111.
"Sunscreen." *Consumer Reports*, Vol. 56, No. 6, June 1991.

How does bleach get clothes white?

Sullivan, Sandy, manager of consumer information and education, and Dave Deleeuw, project leader. The Clorox Company, Oakland, Calif.

How do they get rid of radioactive nuclear waste?

League of Women Voters Educational Fund Staff. *Nuclear Waste Primer*. New York: Nicklyon's Books, 1987.
Murray, Raymond L., and Judith A. Powell. *Understanding Radioactive Waste*. Columbus, Ohio: Battelle Press, 1988.
Office of Civilian Radioactive Waste Management, Department of Energy, 1000 Independence Avenue, Washington, D.C. 20585.

How do they keep Coca-Cola drinkable in polar regions?

Adams, Robert. Consumer Affairs Office, Coca-Cola USA, Atlanta, Ga.; (404) 676-2603.

How do they know if a runner false-starts in an Olympic sprint?

International Amateur Athletic Federation 1990–91 Handbook, 2nd ed. London: International Amateur Athletic Federation.
Mahoney, Peter "Duffy," assistant to the executive director for teams, technical services and training. The Athletics Congress, Indianapolis, Ind.
1987 International Athletic Foundation/IAAF Scientific Project Report on the II World Championships in Athletics.

Rosen, Mel, sprint coach for the U.S. Olympic team in 1984, head coach for the 1992 Olympic track team, head coach at Auburn University, Auburn, Ala.

How do crocodiles clean their teeth?

Mayer, Greg, Postdoctoral Fellow, National Museum of Natural History, Smithsonian Institution, Washington, D.C.
Neill, Wilfred T. *Last of the Ruling Reptiles.* New York: Columbia University Press, 1971.
Taylor, Peter, reptile keeper, and John Behler, Reptile Department curator, New York Zoological Society, Bronx, N.Y.
Whitfield, Dr. Philip, ed. *Macmillan Illustrated Animal Encyclopedia.* New York: Macmillan, 1984.

How do lawyers get paid when a company files for bankruptcy?

Harwood, Mitchel. Gibson, Dunn and Crutcher, (212) 351-4000.

How do they decide how much to pay the queen of England?

People, Royals Special Issue, 1990.
Time, April 27 and December 27, 1971, May 15, 1978, April 6, 1990.
The Washington Post, February 7, 1991.

How do they raise a sunken ship?

Baptist, C. N. *Salvage Operations.* London: Stanford Maritime Press, 1979.
Barranca, Peter S. "Ship Salvaging." *McGraw-Hill Encyclopedia of Science and Technology.* New York: McGraw-Hill, 1987.
Mowat, Farley. *The Grey Seas Under.* Boston: Little, Brown, 1958.
Peterson, Mendel L. *History Under the Sea: A Handbook for*

Underwater Exploration. Washington, D.C.: Smithsonian Institution, 1969.

How do they know how long a second is?

Beehler, Roger, manager. Time and Frequency Broadcast Services, National Institute of Standards and Technology, Boulder, Colo. (NIST is a division of the U.S. Commerce Department.)

How do they decide what to charge for a minute of advertising during the broadcast of the Super Bowl?

Furie, Val, sports planner. CBS, (212) 975-1029.
New York Times, The, January 6, 1991.
Sella, Bill, vice-president. Sports Sales, ABC, (212) 456-7777.

How do they know what a particular dinosaur ate?

Gaffney, Dr. Gene. Department of Paleontology, American Museum of Natural History, New York, N.Y.
Lambert, David. *A Field Guide to Dinosaurs: The First Complete Guide to Every Dinosaur Now Known.* New York: Avon Books, 1983.
Lauber, Patricia. *The News About Dinosaurs.* New York: Bradbury Press, 1989.
Wilford, John Noble. *The Riddle of the Dinosaur.* New York: Alfred A. Knopf, 1985.

How do they measure the heat of distant stars?

Arny, Tom, associate professor of astronomy. University of Massachusetts in Amherst.
Encyclopedia Americana. Danbury, Conn.: Grolier Inc., 1988.
Spitzak, John, Ph.D. student. Department of Physics and Astronomy, University of Massachusetts.

How do they know there's going to be another ice age in two thousand years?

Burten, Bjorn. *The Ice Age*. New York: G. P. Putnam and Sons, 1969.

Chorlton, Windsor. *Ice Ages*. Alexandria, Va.: Time-Life Books, 1983.

Erickson, Jon. *Ice Ages: Past and Future*. Blue Ridge Summit, Penn.: TAB Books, 1990.

Major, Virginia L. Geologic Inquiries Office, U.S. Geologic Survey, Reston, Va., (703) 648-4383.

Radok, Uwe. "The Antarctic Ice." *Scientific American*, August 1985.

?

Index